# Swift Essentials
## *Second Edition*

Discover how to build iOS and watchOS applications in
Swift 2 using Xcode

**Dr Alex Blewitt**

open source
community experience distilled

BIRMINGHAM - MUMBAI

# Swift Essentials
## *Second Edition*

First published: December 2014

Second Edition: January 2016

Production reference: 1200116

Published by Packt Publishing Ltd.
Livery Place
35 Livery Street
Birmingham B3 2PB, UK.

ISBN 978-1-78588-887-8

www.packtpub.com

# Credits

**Author**
Dr Alex Blewitt

**Reviewer**
Antonio Bello

**Commissioning Editor**
Kartikey Pandey

**Acquisition Editor**
Denim Pinto

**Content Development Editor**
Preeti Singh

**Technical Editor**
Siddhesh Patil

**Copy Editor**
Priyanka Ravi

**Project Coordinator**
Milton D'souza

**Proofreader**
Safis Editing

**Indexer**
Hemangini Bari

**Graphics**
Disha Haria

**Production Coordinator**
Arvindkumar Gupta

**Cover Work**
Arvindkumar Gupta

# About the Author

**Dr Alex Blewitt** has over 20 years of experience in Objective-C, and he has been using Apple frameworks since NeXTstep 3.0. He upgraded his NeXTstation for a TiBook when Apple released Mac OS X in 2001, and he has been developing on it ever since.

Alex currently works for an investment bank in London, writes for the online technology news site InfoQ, and has published two other books for Packt Publishing. He also has a number of apps on the Apple AppStore through Bandlem Limited. When he's not working on technology and if the weather is nice, he likes to go flying from the nearby Cranfield airport.

Alex writes regularly at his blog, `http://alblue.bandlem.com`, as well tweeting regularly on Twitter as `@alblue`.

# Acknowledgments

This book would not have been possible without the ongoing love and support of my wife, Amy, who has helped me through both the highs and lows of life. She gave me the freedom to work during the many late nights and weekends that it takes to produce a book and its associated code repository. She truly is the Lem of my life.

I'd also like to thank my parents, Ann and Derek, for their encouragement and support during my formative years. It was this work ethic that allowed me to start my technology career as a teenager and to incorporate my first company before I was 25. I'd also like to congratulate them on their 50th wedding anniversary in 2015, and I look forward to reaching that goal with Amy.

Thanks are due, especially, to the reviewer of this version of the book: Antonio Bello, as well as the previous version of this book: Nate Cook, James Robert, and Arvid Gerstmann, who provided excellent feedback on the contents of this book during development and caught many errors in both the text and code. Any remaining errors are my own.

I'd also like to thank my children Sam and Holly for inspiring me, and I hope that they too can achieve anything that they set their minds to.

Finally, I'd like to thank Ben Moseley, and Eren Kotan, both of whom introduced me to NeXT in the first place and set my career going on a twenty year journey to this book

# About the Reviewer

**Antonio Bello** is a veteran software developer who started writing code when memory was measured in bytes instead of gigabytes and storage was an optional add-on. During his professional career, he's worked with several languages and technologies until he landed on the Apple planet.

Today, he loves developing apps for the iPhone, Apple Watch, Apple TV, and their respective backends. Although he still thinks Objective-C is a great and unconventional language, he prefers and has used Swift ever since it's been announced.

# www.PacktPub.com

## Support files, eBooks, discount offers, and more

For support files and downloads related to your book, please visit www.PacktPub.com.

Did you know that Packt offers eBook versions of every book published, with PDF and ePub files available? You can upgrade to the eBook version at www.PacktPub.com and as a print book customer, you are entitled to a discount on the eBook copy. Get in touch with us at service@packtpub.com for more details.

At www.PacktPub.com, you can also read a collection of free technical articles, sign up for a range of free newsletters and receive exclusive discounts and offers on Packt books and eBooks.

https://www2.packtpub.com/books/subscription/packtlib

Do you need instant solutions to your IT questions? PacktLib is Packt's online digital book library. Here, you can search, access, and read Packt's entire library of books.

## Why subscribe?

- Fully searchable across every book published by Packt
- Copy and paste, print, and bookmark content
- On demand and accessible via a web browser

## Free access for Packt account holders

If you have an account with Packt at www.PacktPub.com, you can use this to access PacktLib today and view nine entirely free books. Simply use your login credentials for immediate access.

# Table of Contents

# Preface

*Swift Essentials* provides an overview of the Swift language and the tooling necessary to write iOS applications. From simple Swift commands on the command line using the open source version of Swift, to interactively testing graphical content on OS X with the Xcode Playground editor, Swift language and syntax is introduced by examples.

This book also introduces end-to-end iOS application development on OS X with Xcode by showing how a simple iOS application can be created, followed by how to use storyboards and custom views to build a more complex networked application.

The book concludes by providing a worked example from scratch that builds up a GitHub repository browser for iOS, along with an Apple Watch application.

## What this book covers

*Chapter 1, Exploring Swift,* presents the open source version of Swift with the Swift Read-Evaluate-Print-Loop (REPL) and introduces the Swift language through examples of standard data types, functions, and looping.

*Chapter 2, Playing with Swift,* demonstrates Swift Xcode Playgrounds as a means to interactively play with Swift code and see graphical results. It also introduces the playground format and shows how playgrounds can be documented.

*Chapter 3, Creating an iOS Swift App,* shows how to create and test an iOS application built in Swift using Xcode, along with an overview of Swift classes, protocols, and enums.

*Chapter 4, Storyboard Applications with Swift and iOS,* introduces the concept of Storyboards as a means of creating a multiscreen iOS application and shows how views in the Interface Builder can be wired to Swift outlets and actions.

*Chapter 5*, *Creating Custom Views in Swift*, covers custom views in Swift using custom table views, laying out nested views, and drawing custom graphics and layered animations.

*Chapter 6*, *Parsing Networked Data*, demonstrates how Swift can talk to networked services using both HTTP and custom stream-based protocols.

*Chapter 7*, *Building a Repository Browser*, uses the techniques described in this book to build a repository browser that can display information about users' GitHub repositories.

*Chapter 8*, *Adding Watch Support*, introduces the capabilities of the Apple Watch and shows how to build an extension for the iOS app to provide data directly on the watch.

The *Appendix*, *References to Swift-related Websites, Blogs, and Notable Twitter Users*, provides additional references and resources to continue learning about Swift.

# What you need for this book

The exercises in this book were written and tested for Swift 2.1, which is bundled with Xcode 7.2, and verified against a development build of Swift 2.2. To experiment with Swift, you will need either a Mac OS X or Linux computer that meets the requirements shown at https://swift.org/download/.

To run the exercises involving Xcode in Chapters 2–8, you need to have a Mac OS X computer running 10.9 or above with Xcode 7.2 or above. If newer versions of Swift are released, check the book's GitHub repository or the book's errata page at Packtpub for details about any changes that may affect the book's content.

> The Swift playground (described in *Chapter 2, Playing with Swift*) is only available as part of Xcode on OS X and is not part of the open source version of Swift.
>
> Also, iOS and watchOS development (*Chapters 3-8*) is only possible on OS X with Xcode; it is not possible to create iOS or watchOS applications on other platforms. Most of the required libraries and modules for iOS development are not available as part of the open source version of Swift.

Xcode can be installed via the App Store as a free download; search for Xcode in the search box. Alternatively, Xcode can be downloaded from https://developer. apple.com/xcode/downloads/, which is referenced from the iOS Developer Center at https://developer.apple.com/devcenter/ios/.

Once Xcode has been installed, it can be launched from `/Applications/Xcode.app` or from Finder. To run the command line-based exercises, Terminal can be launched from `/Applications/Utilities/Terminal.app`, and if Xcode is installed successfully, `swift` can be launched by running `xcrun swift`.

The iOS applications can be developed and tested in the iOS simulator, which comes bundled with Xcode. It is not necessary to have an iOS device to write or test the code. If you want to run the code on your own iOS device, then you will need an Apple ID to sign in, but the application will be limited to directly connected devices. Similarly, the watch application can be tested in a local simulator or on a local device.

Publishing the application to the AppStore requires that you join the **Apple Developer Program**. More information is available at `https://developer.apple.com/programs/`.

# Who this book is for

This book is aimed at developers who are interested in learning the Swift programming language, either using the open source version of Swift on Linux or the version bundled with Xcode on OS X. However, after *Chapter 1, Exploring Swift*, the remainder of the chapters use Xcode features or have iOS examples which can only be used on OS X with Xcode. These chapters show how to write iOS applications on OS X using Swift. No prior programming experience for iOS is assumed, though a basic level of programming experience in a dynamically or statically typed programming language is expected. The reader will be familiar with navigating and using Mac OS X and, in the cases where Terminal commands are required, the developer will have experience of simple shell commands or can pick it up quickly from the examples given.

Developers familiar with Objective-C will know many of the frameworks and libraries mentioned; however, existing knowledge of Objective-C and its frameworks is neither necessary nor assumed.

The sources are provided in a GitHub repository at `https://github.com/alblue/com.packtpub.swift.essentials/`, and they can be used to switch between the content of chapters using the tags in the repository. Knowledge of Git is helpful if you are wanting to navigate between different versions; alternatively, the web-based interface at GitHub may be used instead. It is highly recommended that the reader becomes familiar with Git as it is the standard version control system for Xcode and the de facto standard for open source projects. The reader is invited to read the Git topics at the author's blog `http://alblue.bandlem.com/Tag/git/` if they are unfamiliar and interested in learning more.

# Trademarks

GitHub is a trademark of GitHub Inc., and the examples in this book have not been endorsed, reviewed, or approved by GitHub Inc. Mac and OS X are trademarks of Apple Inc., registered in the U.S. and other countries. iOS is a trademark or registered trademark of Cisco in the U.S. and other countries and is used under license.

# Conventions

In this book, you will find a number of text styles that distinguish between different kinds of information. Here are some examples of these styles and an explanation of their meaning.

Code words in text, database table names, folder names, filenames, file extensions, pathnames, dummy URLs, user input, and Twitter handles are shown as follows: " `"hello".hasPrefix("he")` method compiles and runs successfully on OS X and iOS."

A block of code is set as follows:

```
> var shopping = [ "Milk", "Eggs", "Coffee", ]
shopping: [String] = 3 values {
  [0] = "Milk"
  [1] = "Eggs"
  [2] = "Coffee"
}
```

When we wish to draw your attention to a particular part of a code block, the relevant lines or items are set in bold:

```
func setupView() {
  contentMode = .Redraw
}
```

**New terms** and **important words** are shown in bold. Words that you see on the screen, for example, in menus or dialog boxes, appear in the text like this: "Xcode documentation can be searched by navigating to **Help | Documentation and API Reference**."

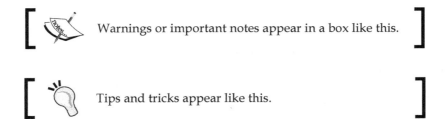

Warnings or important notes appear in a box like this.

Tips and tricks appear like this.

# Reader feedback

Feedback from our readers is always welcome. Let us know what you think about this book—what you liked or disliked. Reader feedback is important for us as it helps us develop titles that you will really get the most out of.

To send us general feedback, simply e-mail feedback@packtpub.com, and mention the book's title in the subject of your message.

If there is a topic that you have expertise in and you are interested in either writing or contributing to a book, see our author guide at www.packtpub.com/authors.

# Customer support

Now that you are the proud owner of a Packt book, we have a number of things to help you to get the most from your purchase.

# Downloading the example code

You can download the example code files from your account at http://www.packtpub.com for all the Packt Publishing books you have purchased. If you purchased this book elsewhere, you can visit http://www.packtpub.com/support and register to have the files e-mailed directly to you.

# Errata

Although we have taken every care to ensure the accuracy of our content, mistakes do happen. If you find a mistake in one of our books—maybe a mistake in the text or the code—we would be grateful if you could report this to us. By doing so, you can save other readers from frustration and help us improve subsequent versions of this book. If you find any errata, please report them by visiting http://www.packtpub.com/submit-errata, selecting your book, clicking on the **Errata Submission Form** link, and entering the details of your errata. Once your errata are verified, your submission will be accepted and the errata will be uploaded to our website or added to any list of existing errata under the Errata section of that title.

To view the previously submitted errata, go to https://www.packtpub.com/books/content/support and enter the name of the book in the search field. The required information will appear under the **Errata** section.

# Piracy

Piracy of copyrighted material on the Internet is an ongoing problem across all media. At Packt, we take the protection of our copyright and licenses very seriously. If you come across any illegal copies of our works in any form on the Internet, please provide us with the location address or website name immediately so that we can pursue a remedy.

Please contact us at copyright@packtpub.com with a link to the suspected pirated material.

We appreciate your help in protecting our authors and our ability to bring you valuable content.

# Questions

If you have a problem with any aspect of this book, you can contact us at questions@packtpub.com, and we will do our best to address the problem.

# 1
# Exploring Swift

Apple announced Swift at WWDC 2014 as a new programming language that combines experience with the Objective-C platform and advances in dynamic and statically typed languages over the last few decades. Before Swift, most code written for iOS and OS X applications was in Objective-C, a set of object-oriented extensions to the C programming language. Swift aims to build upon patterns and frameworks of Objective-C but with a more modern runtime and automatic memory management. In December 2015, Apple open sourced Swift at `https://swift.org` and made binaries available for Linux as well as OS X. The content in this chapter can be run on either Linux or OS X, but the remainder of the book is either Xcode-specific or depends on iOS frameworks that are not open source. Developing iOS applications requires Xcode and OS X.

This chapter will present the following topics:

- How to use the Swift REPL to evaluate Swift code
- The different types of Swift literals
- How to use arrays and dictionaries
- Functions and the different types of function arguments
- Compiling and running Swift from the command line

## Open source Swift

Apple released Swift as an open source project in December 2015, hosted at `https://github.com/apple/swift/` and related repositories. Information about the open source version of Swift is available from the `https://swift.org` site. The open-source version of Swift is similar from a runtime perspective on both Linux and OS X; however, the set of libraries available differ between the two platforms.

For example, the Objective-C runtime was not present in the initial release of Swift for Linux; as a result, several methods that are delegated to Objective-C implementations are not available. `"hello".hasPrefix("he")` compiles and runs successfully on OS X and iOS but is a compile error in the first Swift release for Linux. In addition to missing functions, there is also a different set of modules (frameworks) between the two platforms. The base functionality on OS X and iOS is provided by the `Darwin` module, but on Linux, the base functionality is provided by the `Glibc` module. The `Foundation` module, which provides many of the data types that are outside of the base-collections library, is implemented in Objective-C on OS X and iOS, but on Linux, it is a clean-room reimplementation in Swift. As Swift on Linux evolves, more of this functionality will be filled in, but it is worth testing on both OS X and Linux specifically if cross platform functionality is required.

Finally, although the Swift language and core libraries have been open sourced, this does not apply to the iOS libraries or other functionality in Xcode. As a result, it is not possible to compile iOS or OS X applications from Linux, and building iOS applications and editing user interfaces is something that must be done in Xcode on OS X.

# Getting started with Swift

Swift provides a runtime interpreter that executes statements and expressions. Swift is open source, and precompiled binaries can be downloaded from `https://swift.org/download/` for both OS X and Linux platforms. Ports are in progress to other platforms and operating systems but are not supported by the Swift development team.

The Swift interpreter is called *swift* and on OS X can be launched using the `xcrun` command in a `Terminal.app` shell:

```
$ xcrun swift
Welcome to Swift version 2.2! Type :help for assistance.
>
```

The `xcrun` command allows a toolchain command to be executed; in this case, it finds `/Applications/Xcode.app/Contents/Developer/Toolchains/XcodeDefault.xctoolchain/usr/bin/swift`. The `swift` command sits alongside other compilation tools, such as `clang` and `ld`, and permits multiple versions of the commands and libraries on the same machine without conflicting.

On Linux, the `swift` binary can be executed provided that it and the dependent libraries are in a suitable location.

The Swift prompt displays > for new statements and . for a continuation. Statements and expressions that are typed into the interpreter are evaluated and displayed. Anonymous values are given references so that they can be used subsequently:

```
> "Hello World"
$R0: String = "Hello World"
> 3 + 4
$R1: Int = 7
> $R0
$R2: String = "Hello World"
> $R1
$R3: Int = 7
```

# Numeric literals

Numeric types in Swift can represent both signed and unsigned integral values with sizes of 8, 16, 32, or 64 bits, as well as signed 32 or 64 bit floating point values. Numbers can include underscores to provide better readability; so, 68_040 is the same as 68040:

```
> 3.141
$R0: Double = 3.141
> 299_792_458
$R1: Int = 299792458
> -1
$R2: Int = -1
> 1_800_123456
$R3: Int = 1800123456
```

Numbers can also be written in **binary**, **octal**, or **hexadecimal** using prefixes 0b, 0o (zero and the letter "o") or 0x. Please note that Swift does not inherit C's use of a leading zero (0) to represent an octal value, unlike Java and JavaScript which do. Examples include:

```
> 0b1010011
$R0: Int = 83
> 0o123
$R1: Int = 83
> 0123
$R2: Int = 123
> 0x7b
$R3: Int = 123
```

# Floating point literals

There are three floating point types that are available in Swift which use the IEEE754 floating point standard. The Double type represents 64 bits worth of data, while Float stores 32 bits of data. In addition, Float80 is a specialized type that stores 80 bits worth of data (Float32 and Float64 are available as aliases for Float and Double, respectively, although they are not commonly used in Swift programs).

Some CPUs internally use 80 bit precision to perform math operations, and the Float80 type allows this accuracy to be used in Swift. Not all architectures support Float80 natively, so this should be used sparingly.

By default, floating point values in Swift use the Double type. As floating point representation cannot represent some numbers exactly, some values will be displayed with a rounding error; for example:

```
> 3.141
$R0: Double = 3.141
> Float(3.141)
$R1: Float = 3.1400003
```

Floating point values can be specified in decimal or hexadecimal. Decimal floating point uses e as the exponent for base 10, whereas hexadecimal floating point uses p as the exponent for base 2. A value of AeB has the value $A*10^B$ and a value of 0xApB has the value $A*2^B$. For example:

```
> 299.792458e6
$R0: Double = 299792458
> 299.792_458_e6
$R1: Double = 299792458
> 0x1p8
$R2: Double = 256
> 0x1p10
$R3: Double = 1024
> 0x4p10
$R4: Double = 4096
> 1e-1
$R5: Double = 0.10000000000000001
> 1e-2
$R6: Double = 0.01
> 0x1p-1
$R7: Double = 0.5
> 0x1p-2
$R8: Double = 0.25
> 0xAp-1
$R9: Double = 5
```

# String literals

Strings can contain escaped characters, Unicode characters, and interpolated expressions. Escaped characters start with a slash (\) and can be one of the following:

- \\: This is a literal slash \
- \0: This is the null character
- \' : This is a literal single quote '
- \": This is a literal double quote "
- \t: This is a tab
- \n: This is a line feed
- \r: This is a carriage return
- \u{NNN}: This is a Unicode character, such as the Euro symbol \u{20AC}, or a smiley \u{1F600}

An *interpolated string* has an embedded expression, which is evaluated, converted into a String, and inserted into the result:

```
> "3+4 is \(3+4)"
$R0: String = "3+4 is 7"
> 3+4
$R1: Int = 7
> "7 x 2 is \($R1 * 2)"
$R2: String = "7 x 2 is 14"
```

# Variables and constants

Swift distinguishes between variables (which can be modified) and constants (which cannot be changed after assignment). Identifiers start with an underscore or alphabetic character followed by an underscore or alphanumeric character. In addition, other Unicode character points (such as emoji) can be used although box lines and arrows are not allowed; consult the Swift language guide for the full set of allowable Unicode characters. Generally, Unicode private use areas are not allowed, and identifiers cannot start with a combining character (such as an accent).

Variables are defined with the var keyword, and constants are defined with the let keyword. If the type is not specified, it is automatically inferred:

```
> let pi = 3.141
pi: Double = 3.141
> pi = 3

error: cannot assign to value: 'pi' is a 'let' constant
note: change 'let' to 'var' to make it mutable
```

```
> var i = 0
i: Int = 0
> ++i
$R0: Int = 1
```

Types can be explicitly specified. For example, to store a 32 bit floating point value, the variable can be explicitly defined as a `Float`:

```
> let e:Float = 2.718
e: Float = 2.71799994
```

Similarly, to store a value as an unsigned 8 bit integer, explicitly declare the type as `UInt8`:

```
> let ff:UInt8 = 255
ff: UInt8 = 255
```

A number can be converted to a different type using the type initializer or a literal that is assigned to a variable of a different type, provided that it does not underflow or overflow:

```
> let ooff = UInt16(ff)
ooff: UInt16 = 255
> Int8(255)
error: integer overflows when converted from 'Int' to 'Int8'
Int8(255)
^
> UInt8(Int8(-1))
error: negative integer cannot be converted to unsigned type 'UInt8'
UInt8(Int8(-1))
^
```

# Collection types

Swift has three collection types: *Array*, *Dictionary*, and *Set*. They are strongly typed and generic, which ensures that the values of types that are assigned are compatible with the element type. Collections that are defined with `var` are mutable; collections defined with `let` are immutable.

The literal syntax for arrays uses `[]` to store a comma-separated list:

```
> var shopping = [ "Milk", "Eggs", "Coffee", ]
shopping: [String] = 3 values {
  [0] = "Milk"
  [1] = "Eggs"
  [2] = "Coffee"
}
```

Literal dictionaries are defined with a comma-separated `[key:value]` format for entries:

```
> var costs = [ "Milk":1, "Eggs":2, "Coffee":3, ]
costs: [String : Int] = {
  [0] = { key = "Coffee" value = 3 }
  [1] = { key = "Milk"   value = 1 }
  [2] = { key = "Eggs"   value = 2 }
}
```

For readability, array and dictionary literals can have a trailing comma. This allows initialization to be split over multiple lines, and if the last element ends with a trailing comma, adding new items does not result in an SCM diff to the previous line.

Arrays and dictionaries can be indexed using subscript operators that are reassigned and added to:

```
> shopping[0]
$R0: String = "Milk"
> costs["Milk"]
$R1: Int? = 1
> shopping.count
$R2: Int = 3
> shopping += ["Tea"]
> shopping.count
$R3: Int = 4
> costs.count
$R4: Int = 3
> costs["Tea"] = "String"
error: cannot assign a value of type 'String' to a value of type
'Int?'
> costs["Tea"] = 4
> costs.count
$R5: Int = 4
```

Sets are similar to dictionaries; the keys are unordered and can be looked up efficiently. However, unlike dictionaries, keys don't have an associated value. As a result, they don't have array subscripts, but they do have the `insert`, `remove`, and `contains` methods. They also have efficient set intersection methods, such as `union` and `intersect`. They can be created from an array literal if the type is defined or using the set initializer directly:

```
> var shoppingSet: Set = [ "Milk", "Eggs", "Coffee", ]
> // same as: shoppingSet = Set( [ "Milk", "Eggs", "Coffee", ] )
```

```
> shoppingSet.contains("Milk")
$R6: Bool = true
> shoppingSet.contains("Tea")
$R7: Bool = false
> shoppingSet.remove("Coffee")
$R8: String? = "Coffee"
> shoppingSet.remove("Tea")
$R9: String? = nil
> shoppingSet.insert("Tea")
> shoppingSet.contains("Tea")
$R10: Bool = true
```

 When creating sets, use the explicit Set constructor as otherwise the type will be inferred to be an Array, which will have a different performance profile.

# Optional types

In the previous example, the return type of costs["Milk"] is Int? and not Int. This is an *optional type*; there may be an Int value or it may be empty. For a dictionary containing elements of type T, subscripting the dictionary will have an Optional<T> type, which can be abbreviated as T? If the value doesn't exist in the dictionary, then the returned value will be nil. Other object-oriented languages, such as Objective-C, C++, Java, and C#, have optional types by default; any object value (or pointer) can be null. By representing optionality in the type system, Swift can determine whether a value really has to exist or might be nil:

```
> var cannotBeNil: Int = 1
cannotBeNil: Int = 1
> cannotBeNil = nil
error: cannot assign a value of type 'nil' to a value of type 'Int'
cannotBeNil = nil
              ^
> var canBeNil: Int? = 1
canBeNil: Int? = 1
> canBeNil = nil
$R0: Int? = nil
```

Optional types can be explicitly created using the Optional constructor. Given a value x of type X, an optional X? value can be created using Optional(x). The value can be tested against nil to find out whether it contains a value and then unwrapped with opt!, for example:

```
> var opt = Optional(1)
opt: Int? = 1
> opt == nil
$R1: Bool = false
> opt!
$R2: Int = 1
```

If a nil value is unwrapped, an error occurs:

```
> opt = nil
> opt!
fatal error: unexpectedly found nil while unwrapping an Optional value
Execution interrupted. Enter Swift code to recover and continue.
Enter LLDB commands to investigate (type :help for assistance.)
```

Particularly when working with Objective-C based APIs, it is common for values to be declared as an optional although they are always expected to return a value. It is possible to declare such variables as *implicitly unwrapped optionals*; these variables behave as optional values (they may contain nil), but when the value is accessed, they are automatically unwrapped on demand:

```
> var implicitlyUnwrappedOptional:Int! = 1
implicitlyUnwrappedOptional: Int! = 1
> implicitlyUnwrappedOptional + 2
3
> implicitlyUnwrappedOptional = nil
> implicitlyUnwrappedOptional + 2
fatal error: unexpectedly found nil while unwrapping an Optional value
```

In general, implicitly unwrapped optionals should be avoided as they are likely to lead to errors. They are mainly useful for interaction with existing Objective-C APIs when the value is known to have an instance.

# Nil coalescing operator

Swift has a *nil coalescing operator*, which is similar to Groovy's `?:` operator or C#'s `??` operator. This provides a means to specify a default value if an expression is `nil`:

```
> 1 ?? 2
$R0: Int = 1
> nil ?? 2
$R1: Int = 2
```

The `nil` coalescing operator can also be used to unwrap an optional value. If the optional value is present, it is unwrapped and returned; if it is missing, then the right-hand side of the expression is returned. Similar to the `||` shortcut, and the `&&` operators, the right-hand side is not evaluated unless necessary:

```
> costs["Tea"] ?? 0
$R2: Int = 4
> costs["Sugar"] ?? 0
$R3: Int = 0
```

# Conditional logic

There are three key types of conditional logic in Swift (known as branch statements in the grammar): the `if` statement, the `switch` statement, and the `guard` statement. Unlike other languages, the body of the `if` must be surrounded with braces `{}`; and if typed in at the interpreter, the `{` opening brace must be on the same line as the `if` statement. The `guard` statement is a specialized `if` statement for use with functions and is covered in the section on functions later in this chapter.

# If statements

Conditionally unwrapping an optional value is so common that a specific Swift pattern *optional binding* has been created to avoid evaluating the expression twice:

```
> var shopping = [ "Milk", "Eggs", "Coffee", "Tea", ]
> var costs = [ "Milk":1, "Eggs":2, "Coffee":3, "Tea":4, ]
> var cost = 0
> if let cc = costs["Coffee"] {
.    cost += cc
. }
> cost
$R0: Int = 3
```

The `if` block only executes if the optional value exists. The definition of the `cc` constant only exists for the body of the `if` block, and it does not exist outside of that scope. Furthermore, `cc` is a non-optional type, so it is guaranteed not to be `nil`.

Swift 1 only allowed a single `let` assignment in an `if` block causing a pyramid of nested `if` statements. Swift 2 allows multiple comma-separated `let` assignments in a single `if` statement.

```
> if let cm = costs["Milk"], let ct =
costs["Tea"] {
.   cost += cm + ct
. }
> cost
$R1: Int = 8
```

To execute an alternative block if the item cannot be found, an `else` block can be used:

```
> if let cb = costs["Bread"] {
.   cost += cb
. } else {
.   print("Cannot find any Bread")
. }
Cannot find any Bread
```

Other boolean expressions can include the `true` and `false` literals, and any expression that conforms to the `BooleanType` protocol, the `==` and `!=` equality operators, the `===` and `!==` identity operators, as well as the `<`, `<=`, `>`, and `>=` comparison operators. The `is` type operator provides a test to see whether an element is of a particular type.

The difference between the equality operator and the identity operator is relevant for classes or other reference types. The equality operator asks *Are these two values equivalent to each other?*, whereas the identity operator asks *Are these two references equal to each other?*

There is a boolean operator that is specific to Swift, which is the `~=` *pattern match operator*. Despite the name, this isn't anything to do with regular expressions; rather, it's a way of asking whether a pattern matches a particular value. This is used in the implementation of the `switch` block, which is covered in the next section.

As well as the `if` statement, there is a *ternary if expression* that is similar to other languages. After a condition, a question mark (?) is used followed by an expression to be used if the condition is true, then a colon (:) followed by the false expression:

```
> var i = 17
i: Int = 17
> i % 2 == 0 ? "Even" : "Odd"
$R0: String = "Odd"
```

# Switch statements

Swift has a `switch` statement that is similar to C and Java's `switch`. However, it differs in two important ways. Firstly, `case` statements no longer have a default fall-through behavior (so there are no bugs introduced by missing a `break` statement), and secondly, the value of the `case` statements can be expressions instead of values, pattern matching on type and range. At the end of the corresponding `case` statement, the evaluation jumps to the end of the `switch` block unless the `fallthrough` keyword is used. If no `case` statements match, the `default` statements are executed.

A `default` statement is required when the list of cases is not exhaustive. If they are not, the compiler will give an error saying that the list is not exhaustive and that a `default` statement is required.

```
> var position = 21
position: Int = 21
> switch position {
.    case 1: print("First")
.    case 2: print("Second")
.    case 3: print("Third")
.    case 4...20: print("\(position)th")
.    case position where (position % 10) == 1:
.      print("\(position)st")
.    case let p where (p % 10) == 2:
.      print("\(p)nd")
.    case let p where (p % 10) == 3:
.      print("\(p)rd")
.    default: print("\(position)th")
. }
21st
```

In the preceding example, the expression prints out First, Second, or Third if the position is 1, 2, or 3, respectively. For numbers between 4 and 20 (inclusive), it prints out the position with a th ordinal. Otherwise, for numbers that end with 1, it prints st; for numbers that end with 2, it prints nd, and for numbers that end with 3, it prints rd. For all other numbers it prints th.

The 4...20 range expression in a case statement represents a pattern. If the value of the expression matches that pattern, then the corresponding statements will be executed:

```
> 4...10 ~= 4
$R0: Bool = true
> 4...10 ~= 21
$R1: Bool = false
```

There are two range operators in Swift: an inclusive or *closed range*, and an exclusive or *half-open range*. The closed range is specified with three dots; so 1...12 will give a list of integers between one and twelve. The half-open range is specified with two dots and a less than operator; so 1..<10 will provide integers from 1 to 9 but excluding 10.

The where clause in the switch block allows an arbitrary expression to be evaluated provided that the pattern matches. These are evaluated in order, in the sequence they are in the source file. If a where clause evaluates to true, then the corresponding set of statements will be executed.

The let variable syntax can be used to define a constant that refers to the value in the switch block. This local constant can be used in the where clause or the corresponding statements for that specific case. Alternatively, variables can be used from the surrounding scope.

If multiple case statements need to match the same pattern, they can be separated with commas as an expression list. Alternatively, the fallthrough keyword can be used to allow the same implementation to be used for multiple case statements.

# Iteration

Ranges can be used to iterate a fixed number of times, for example, for i in 1...12. To print out these numbers, a loop such as the following can be used:

```
> for i in 1...12 {
.    print("i is \(i)")
. }
```

If the number is not required, then an underscore (_) can be used as a hole to act as a throwaway value. An underscore can be assigned to but not read:

```
> for _ in 1...12 {
.    print("Looping...")
. }
```

However, it is more common to iterate over a collection's contents using a `for in` pattern. This steps through each of the items in the collection, and the body of the `for` loop is executed over each one:

```
> var shopping = [ "Milk", "Eggs", "Coffee", "Tea", ]
> var costs = [ "Milk":1, "Eggs":2, "Coffee":3, "Tea":4, ]
> var cost = 0
> for item in shopping {
.    if let itemCost = costs[item] {
.        cost += itemCost
.    }
. }
> cost
cost: Int = 10
```

To iterate over a dictionary, it is possible to extract the keys or the values and process them as an array:

```
> Array(costs.keys)
$R0: [String] = 4 values {
   [0] = "Coffee"
   [1] = "Milk"
   [2] = "Eggs"
   [3] = "Tea"
}
> Array(costs.values)
$R1: [Int] = 4 values {
   [0] = 3
   [1] = 1
   [2] = 2
   [3] = 4
}
```

 The order of keys in a dictionary is not guaranteed; as the dictionary changes, the order may change.

Converting a dictionary's values to an array will result in a copy of the data being made, which can lead to poor performance. As the underlying `keys` and `values` are of a `LazyMapCollection` type, they can be iterated over directly:

```
> costs.keys
$R2: LazyMapCollection<[String : Int], String> = {
  _base = {
    _base = 4 key/value pairs {
      [0] = { key = "Coffee" value = 3 }
      [1] = { key = "Milk"   value = 1 }
      [2] = { key = "Eggs"   value = 2 }
      [3] = { key = "Tea"    value = 4 }
    }
  _transform =
  }
}
```

To print out all the keys in a dictionary, the `keys` property can be used with a `for in` loop:

```
> for item in costs.keys {
.    print(item)
. }
Coffee
Milk
Eggs
Tea
```

# Iterating over keys and values in a dictionary

Traversing a dictionary to obtain all of the keys and then subsequently looking up values will result in searching the data structure twice. Instead, both the key and the value can be iterated at the same time, using a *tuple*. A tuple is like a fixed-sized array, but one that allows assigning pairs (or more) of values at a time:

```
> var (a,b) = (1,2)
a: Int = 1
b: Int = 2
```

Tuples can be used to iterate pairwise over both the keys and values of a dictionary:

```
> for (item,cost) in costs {
.    print("The \(item) costs \(cost)")
. }
The Coffee costs 3
The Milk costs 1
The Eggs costs 2
The Tea costs 4
```

Both `Array` and `Dictionary` conform to the `SequenceType` protocol, which allows them to be iterated with a `for in` loop. Collections (as well as other objects, such as `Range`) that implement `SequenceType` have a `generate` method, which returns a `GeneratorType` that allows the data to be iterated over. It is possible for custom Swift objects to implement `SequenceType` to allow them to be used in a `for in` loop.

# Iteration with for loops

Although the most common use of the `for` operator in Swift is in a `for in` loop, it is also possible (in Swift 1 and 2) to use a more traditional form of the `for` loop. This has an initialization, a condition that is tested at the start of each loop, and a step operation that is evaluated at the end of each loop. Although the parentheses around the `for` loop are optional, the braces for the block of code are mandatory.

 It has been proposed that both the traditional `for` loop and the increment/decrement operators should be removed from Swift 3. It is recommended that these forms of loops be avoided where possible.

Calculating the sum of integers between 1 and 10 can be performed without using the range operator:

```
> var sum = 0
. for var i=0; i<=10; ++i {
.     sum += i
. }
sum: Int = 55
```

If multiple variables need to be updated in the `for` loop, Swift has an *expression list* that is a set of comma-separated expressions. To step through two sets of variables in a for loop, the following can be used:

```
> for var i = 0,j = 10; i<=10 && j >= 0; ++i,--j {
.     print("\(i), \(j)")
. }
0, 10
1, 9
...
9, 1
10, 0
```

Apple recommends the use of ++i instead of i++ (and conversely, --i instead of i--) because they will return the result of i after the operation, which may be the expected value. As noted earlier, these operators may be removed in a future version of Swift.

# Break and continue

The break statement leaves the innermost loop early, and control jumps to the end of the loop. The continue statement takes execution to the top of the innermost loop and the next item.

To *break* or *continue* from nested loops, a *label* can be used. Labels in Swift can only be applied to a loop statement, such as while or for. A label is introduced by an identifier and a colon just before the loop statement:

```
> var deck = [1...13, 1...13, 1...13, 1...13]
> suits: for suit in deck {
.    for card in suit {
.      if card == 3 {
.        continue // go to next card in same suit
.      }
.      if card == 5 {
.        continue suits // go to next suit
.      }
.      if card == 7 {
.        break // leave card loop
.      }
.      if card == 13 {
.        break suits // leave suit loop
.      }
.    }
. }
```

# Functions

Functions can be created using the func keyword, which takes a set of arguments and a body of statements. The return statement can be used to leave a function:

```
> var shopping = [ "Milk", "Eggs", "Coffee", "Tea", ]
> var costs = [ "Milk":1, "Eggs":2, "Coffee":3, "Tea":4, ]
> func costOf(items:[String], _ costs:[String:Int]) -> Int {
.    var cost = 0
```

```
.    for item in items {
.      if let ci = costs[item] {
.        cost += ci
.      }
.    }
.    return cost
.  }
> costOf(shopping,costs)
$R0: Int = 10
```

The return type of the function is specified after the arguments with an arrow (->). If missing, the function cannot return a value; if present, the function must return a value of that type.

The underscore (_) on the front of the costs parameter is required to avoid it being a named argument. The second and subsequent arguments in Swift functions are implicitly named. To ensure that it is treated as a positional argument, the _ before the argument name is required.

Functions with *positional arguments* can be called with parentheses, such as the costOf(shopping,costs) call. If a function takes no arguments, then the parentheses are still required.

The foo() expression calls the foo function with no argument. The foo expression represents the function itself, so an expression, such as let copyOfFoo = foo, results in a copy of the function; as a result, copyOfFoo() and foo() have the same effect.

# Named arguments

Swift also supports *named arguments,* which can either use the name of the variable or can be defined with an *external parameter name.* To modify the function to support calling with basket and prices as argument names, the following can be done:

```
> func costOf(basket items:[String], prices costs:[String:Int]) -> Int
{
.    var cost = 0
.    for item in items {
.      if let ci = costs[item] {
.        cost += ci
.      }
.    }
.    return cost
.  }
> costOf(basket:shopping, prices:costs)
$R1: Int = 10
```

This example defines external parameter names `basket` and `prices` for the function. The function signature is often referred to as `costOf(basket:prices:)` and is useful when it may not be clear what the arguments are for (particularly if they are of the same type).

# Optional arguments and default values

Swift functions can have *optional arguments* by specifying *default values* in the function definition. When the function is called, if an optional argument is missing, the default value for that argument is used.

> An optional argument is one that can be omitted in the function call rather than a required argument that takes an optional value. This naming is unfortunate. It may help to think of these as default arguments rather than optional arguments.

A default parameter value is specified after the type in the function signature, with an equals (=) and then the expression. This expression is re-evaluated each time the function is called without a corresponding argument.

In the `costOf` example, instead of passing the value of `costs` each time, it could be defined with a default parameter:

```
> func costOf(items items:[String], costs:[String:Int] = costs) -> Int
{
.    var cost = 0
.    for item in items {
.      if let ci = costs[item] {
.        cost += ci
.      }
.    }
.    return cost
. }
> costOf(items:shopping)
$R2: Int = 10
> costOf(items:shopping, costs:costs)
$R3: Int = 10
```

Please note that the captured `costs` variable is bound when the function is defined.

> To use a named argument as the first parameter in a function, the argument name has to be duplicated. Swift 1 used a hash (#) to represent an implicit parameter name, but this was removed from Swift 2.

# Guards

It is a common code pattern for a function to require arguments that meet certain conditions before the function can run successfully. For example, an optional value must have a value or an integer argument must be in a certain range.

Typically, the pattern to implement this is either to have a number of `if` statements that break out of the function at the top, or to have an `if` block wrapping the entire method body:

```
if card < 1 || card > 13 {
  // report error
  return
}

// or alternatively:

if card >= 1 && card <= 13 {
  // do something with card
} else {
  // report error
}
```

Both of these approaches have drawbacks. In the first case, the condition has been negated; instead of looking for valid values, it's checking for invalid values. This can cause subtle bugs to creep in; for example, `card < 1 && card > 13` would never succeed, but it may inadvertently pass a code review. There's also the problem of what happens if the block doesn't `return` or `break`; it could be perfectly valid Swift code but still include errors.

In the second case, the main body of the function is indented at least one level in the body of the `if` statement. When multiple conditions are required, there may be many nested `if` statements, each with their own error handling or cleanup requirements. If new conditions are required, then the body of the code may be indented even further, leading to code churn in the repository even when only whitespace has changed.

Swift 2 adds a `guard` statement, which is conceptually identical to an `if` statement, except that it only has an `else` clause body. In addition, the compiler checks that the `else` block returns from the function, either by returning or by throwing an exception:

```
> func cardName(value:Int) -> String {
.   guard value >= 1 && value <= 13 else {
.     return "Unknown card"
.   }
```

```
  .     let cardNames = [11:"Jack",12:"Queen",13:"King",1:"Ace",]
  .     return cardNames[value] ?? "\(value)"
  . }
```

The Swift compiler checks that the guard else block leaves the function, and reports a compile error if it does not. Code that appears after the guard statement can guarantee that the value is in the 1...13 range without having to perform further tests.

The guard block can also be used to perform *optional binding*; if the guard condition is a let assignment that performs an optional test, then the code that is subsequent to the guard statement can use the value without further unwrapping:

```
> func firstElement(list:[Int]) -> String {
.     guard let first = list.first else {
.       return "List is empty"
.     }
.     return "Value is \(first)"
. }
```

As the first element of an array is an optional value, the guard test here acquires the value and unwraps it. When it is used later in the function, the unwrapped value is available for use without requiring further unwrapping.

# Multiple return values and arguments

So far, the examples of functions have all returned a single type. What happens if there is more than one return result from a function? In an object-oriented language, the answer is to return a class; however, Swift has tuples, which can be used to return multiple values. The type of a tuple is the type of its constituent parts:

```
> var pair = (1,2)
pair: (Int, Int) ...
```

This can be used to return multiple values from the function; instead of just returning one value, it is possible to return a tuple of values.

 Swift also has in-out arguments, which will be seen in the *Handling errors* section of *Chapter 6, Parsing Networked Data*.

Separately, it is also possible to take a variable number of arguments. A function can easily take an array of values with [], but Swift provides a mechanism to allow calling with multiple arguments, using a *variadic* parameter, which is denoted as an ellipses (…) after the type. The value can then be used as an array in the function.

Swift 1 only allowed the variadic argument as the last argument; Swift 2 relaxed that restriction to allow a single variadic argument to appear anywhere in the function's parameters.

Taken together, these two features allow the creation of a `minmax` function, which returns both the minimum and maximum from a list of integers:

```
> func minmax(numbers:Int…) -> (Int,Int) {
.     var min = Int.max
.     var max = Int.min
.     for number in numbers {
.       if number < min {
.         min = number
.       }
.       if number > max {
.         max = number
.       }
.     }
.     return (min,max)
. }
> minmax(1,2,3,4)
$R0: (Int, Int) = {
  0 = 1
  1 = 4
}
```

The `numbers:Int…` argument indicates that a variable number of arguments can be passed into the function. Inside the function, it is processed as an ordinary array; in this case, iterating through using a `for in` loop.

 `Int.max` is a constant representing the largest `Int` value, and `Int.min` is a constant representing the smallest `Int` value. Similar constants exist for other integral types, such as `UInt8.max`, and `Int64.min`.

What if no arguments are passed in? If run on a 64 bit system, then the output will be:

```
> minmax()
$R1: (Int, Int) = {
  0 = 9223372036854775807
  1 = -9223372036854775808
}
```

This may not make sense for a `minmax` function. Instead of returning an error value or a default value, the type system can be used. By making the tuple optional, it is possible to return a `nil` value if it doesn't exist, or a tuple if it does:

```
> func minmax(numbers:Int...) -> (Int,Int)? {
.    var min = Int.max
.    var max = Int.min
.    if numbers.count == 0 {
.      return nil
.    } else {
.      for number in numbers {
.        if number < min {
.          min = number
.        }
.        if number > max {
.          max = number
.        }
.      }
.      return(min,max)
.    }
. }
> minmax()
$R2: (Int, Int)? = nil
> minmax(1,2,3,4)
$R3: (Int, Int)? = (0 = 1, 1 = 4)
> var (minimum,maximum) = minmax(1,2,3,4)!
minimum: Int = 1
maximum: Int = 4
```

Returning an optional value allows the caller to determine what should happen in cases where the maximum and minimum are not present.

> If a function does not always have a valid return value, use an optional type to encode that possibility into the type system.

# Returning structured values

A tuple is an ordered set of data. The entries in the tuple are ordered, but it can quickly become unclear as to what data is stored, particularly if they are of the same type. In the `minmax` tuple, it is not clear which value is the minimum and which value is the maximum, and this can lead to subtle programming errors later on.

A structure (`struct`) is like a tuple but with named values. This allows members to be accessed by name instead of by position, leading to fewer errors and greater transparency. Named values can be added to tuples as well; in essence, tuples with named values are anonymous structures.

 Structs are passed in a copy-by-value manner like tuples. If two variables are assigned the same struct or tuple, then changes to one do not affect the values of another.

A `struct` is defined with the `struct` keyword and has variables or values in the body:

```
> struct MinMax {
.    var min:Int
.    var max:Int
. }
```

This defines a `MinMax` type, which can be used in place of any of the types that are seen so far. It can be used in the `minmax` function to return a `struct` instead of a tuple:

```
> func minmax(numbers:Int...) -> MinMax? {
.    var minmax = MinMax(min:Int.max, max:Int.min)
.    if numbers.count == 0 {
.      return nil
.    } else {
.      for number in numbers {
.        if number < minmax.min {
.          minmax.min = number
.        }
.        if number > minmax.max {
.          minmax.max = number
.        }
.      }
.      return minmax
.    }
. }
```

The `struct` is initialized with a type initializer; if `MinMax()` is used, then the default values for each of the structure types are given (based on the structure definition), but these can be overridden explicitly if desired with `MinMax(min:-10,max:11)`. For example, if the `MinMax` struct is defined as `struct MinMax { var min:Int = Int.max; var max:Int = Int.min }`, then `MinMax()` will return a structure with the appropriate minimum and maximum values filled in.

When a structure is initialized, all the non-optional fields must be assigned. They can be passed in as named arguments in the initializer or specified in the structure definition.

Swift also has classes; these are covered in the Swift classes section in the next chapter.

# Error handling

In the original Swift release, error handling consisted of either returning a `Bool` or an optional value from function results. This tended to work inconsistently with Objective-C, which used an optional `NSError` pointer on various calls that was set if a condition had occurred.

Swift 2 adds an exception-like error model, which allows code to be written in a more compact way while ensuring that errors are handled accordingly. Although this isn't implemented in quite the same way as C++ exception handling, the semantics of the error handling are quite similar.

Errors can be created using a new `throw` keyword, and errors are stored as a subtype of `ErrorType`. Although swift `enum` values (covered in *Chapter 3, Creating an iOS Swift App*) are often used as error types, `struct` values can be used as well.

Exception types can be created as subtypes of `ErrorType` by appending the supertype after the type name:

```
> struct Oops:ErrorType {
.    let message:String
. }
```

Exceptions are thrown using the `throw` keyword and creating an instance of the exception type:

```
> throw Oops(message:"Something went wrong")
$E0: Oops = {
  message = "Something went wrong"
}
```

 The REPL displays exception results with the $E prefix; ordinary results are displayed with the $R prefix.

# Throwing errors

Functions can declare that they return an error using the `throws` keyword before the return type, if any. The previous `cardName` function, which returned a dummy value if the argument was out of range, can be upgraded to throw an exception instead by adding the `throws` keyword before the return type and changing the `return` to a `throw`:

```
> func cardName(value:Int) throws -> String {
.    guard value >= 1 && value <= 13 else {
.       throw Oops(message:"Unknown card")
.    }
.    let cardNames = [11:"Jack",12:"Queen",13:"King",1:"Ace",]
.    return cardNames[value] ?? "\(value)"
. }
```

When the function is called with a real value, the result is returned; when it is passed an invalid value, an exception is thrown instead:

```
> cardName(1)
$R1: String = "Ace"
> cardName(15)
$E2: Oops = {
  message = "Unknown card"
}
```

When interfacing with Objective-C code, methods that take an `NSError**` argument are automatically represented in Swift as methods that throw. In general, any method whose arguments ends in `NSError**` is treated as throwing an exception in Swift.

Exception throwing in C++ and Objective-C is not as performant as exception handling in Swift because the latter does not perform stack unwinding. As a result, exception throwing in Swift is equivalent (from a performance perspective) to dealing with return values. Expect the Swift library to evolve in the future towards a throws-based means of error detection and away from Objective-C's use of `**NSError` pointers.

# Catching errors

The other half of exception handling is the ability to catch errors when they occur. As with other languages, Swift now has a `try/catch` block that can be used to handle error conditions. Unlike other languages, the syntax is a little different; instead of a `try/catch` block, there is a `do/catch` block, and each expression that may throw an error is annotated with its own `try` statement:

```
> do {
.    let name = try cardName(15)
.    print("You chose \(name)")
. } catch {
.    print("You chose an invalid card")
. }
```

When the preceding code is executed, it will print out the generic error message. If a different choice is given, then it will run the successful path instead.

It's possible to capture the error object and use it in the catch block:

```
. } catch let e {
.    print("There was a problem \(e)")
. }
```

 The default `catch` block will bind to a variable called `error` if not specified

Both of these two preceding examples will catch any errors thrown from the body of the code.

 It's possible to catch explicitly based on type if the type is an `enum` that is using pattern matching, for example, `catch Oops(let message)`. However, as this does not work for struct values, it cannot be tested here. *Chapter 3, Creating an iOS Swift App* introduces `enum` types.

Sometimes code will always work, and there is no way it can fail. In these cases, it's cumbersome to have to wrap the code with a `do/try/catch` block when it is known that the problem can never occur. Swift provides a short-cut for this using the `try!` statement, which catches and filters the exception:

```
> let ace = try! cardName(1)
ace: String = "Ace"
```

If the expression really does fail, then it translates to a runtime error and halts the program:

```
> let unknown = try! cardName(15)

Fatal error: 'try!' expression unexpectedly raised an error:
Oops(message: "Unknown card")
```

 Using try! is not generally recommended; if an error occurs then the program will crash. However, it is often used with user interface codes as Objective-C has a number of optional methods and values that are conventionally known not to be nil, such as the reference to the enclosing window.

A better approach is to use try?, which translates the expression into an optional value: if evaluation succeeds, then it returns an optional with a value; if evaluation fails, then it returns a nil value:

```
> let ace = try? cardName(1)
ace: String? = "Ace"
> let unknown = try? cardName(15)
unknown: String? = nil
```

This is handy for use in the if let or guard let constructs, to avoid having to wrap in a do/catch block:

```
> if let card = try? cardName(value) {
.    print("You chose: \(card)")
. }
```

# Cleaning up after errors

It is common to have a function that needs to perform some cleanup before the function returns, regardless of whether the function has completed successfully or not. An example would be working with files; at the start of the function the file may be opened, and by the end of the function it should be closed again, whether or not an error occurs.

A traditional way of handling this is to use an optional value to hold the file reference, and at the end of the method if it is not nil, then the file is closed. However, if there is the possibility of an error occurring during the method's execution, there needs to be a do/catch block to ensure that the cleanup is correctly called, or a set of nested if statements that are only executed if the file is successful.

The downside with this approach is that the actual body of the code tends to be indented several times each with different levels of error handling and recovery at the end of the method. The syntactic separation between where the resource is acquired and where the resource is cleaned up can lead to bugs.

Swift has a `defer` statement, which can be used to register a block of code to be run at the end of the function call. This block is run regardless of whether the function returns normally (with the `return` statement) or if an error occurs (with the `throw` statement). Deferred blocks are executed in reverse order of execution, for example:

```
> func deferExample() {
.    defer { print("C") }
.    print("A")
.    defer { print("B") }
. }
> deferExample()
A
B
C
```

Please note that if a `defer` statement is not executed, then the block is not executed at the end of the method. This allows a `guard` statement to leave the function early, while executing the `defer` statements that have been added so far:

```
> func deferEarly() {
.    defer { print("C") }
.    print("A")
.    return
.    defer { print("B") } // not executed
. }
> deferEarly()
A
C
```

# Command-line Swift

As Swift can be interpreted, it is possible to use it in shell scripts. By setting the interpreter to `swift` with a *hashbang*, the script can be executed without requiring a separate compilation step. Alternatively, Swift scripts can be compiled to a native executable that can be run without the overhead of the interpreter.

# Interpreted Swift scripts

Save the following as `hello.swift`:

```
#!/usr/bin/env xcrun swift
print("Hello World")
```

 In Linux, the first line should point to the location of the `swift` executable, such as `#!/usr/bin/swift`.

After saving, make the file executable by running `chmod a+x hello.swift`. The program can then be run by typing `./hello.swift`, and the traditional greeting will be seen:

```
Hello World
```

Arguments can be passed from the command line and interrogated in the process using the `Process` class through the `arguments` constant. As with other Unix commands, the first element (0) is the name of the process executable; the arguments that are passed from the command line start from one (1).

The program can be terminated using the `exit` function; however, this is defined in the operating system libraries and so it needs to be imported in order to call this function. Modules in Swift correspond to Frameworks in Objective-C and give access to all functions that are defined as public API in the module. The syntax to import all elements from a module is `import module` although it's also possible to import a single function using `import func module.functionName`.

 Not all foundation libraries are implemented for Linux, which results in some differences of behavior. In addition, the underlying module for the base functionality is `Darwin` on iOS and OS X, and is `Glibc` on Linux. These can also be accessed with `import Foundation`, which will include the appropriate operating system module.

A Swift program to print arguments in uppercase can be implemented as a script:

```
#!/usr/bin/env xcrun swift
import func Darwin.exit
# import func Glibc.exit # for Linux
let args = Process.arguments[1..<Process.arguments.count]
for arg in args {
  print("\(arg.uppercaseString)")
}
exit(0)
```

Running this with `hello world` results in the following:

```
$ ./upper.swift hello world
HELLO
WORLD
```

Conventionally, the entry point to Swift programs is via a script called `main.swift`. If starting a Swift-based command-line application project in Xcode, a `main.swift` file will be created automatically. Scripts do not need to have a `.swift` extension; for example, the previous example could be called `upper` and it would still work.

# Compiled Swift scripts

While interpreted Swift scripts are useful for experimenting and writing, each time the script is started, it is interpreted using the Swift compiler and then executed. For simple scripts (such as converting arguments to upper case), this can be a large proportion of the script's execution time.

To compile a Swift script into a native executable, use the `swiftc` command with the `-o` output flag to specify a file to write to. This will then generate an executable that does exactly the same as the interpreted script, only much faster. The `time` command can be used to compare the running time of the interpreted and compiled versions:

```
$ time ./upper.swift hello world     # Interpreted
HELLO
WORLD
real  0m0.145s
$ xcrun swiftc -o upper upper.swift # Compile step
$ time ./upper hello world          # Compiled
HELLO
WORLD
real  0m0.012s
```

Of course, the numbers will vary, and the initial step only happens once, but startup is very lightweight in Swift. The numbers are not meant to be taken in magnitude but rather as relative to each other.

The compile step can also be used to link together many individual Swift files into one executable, which helps create a more organized project; Xcode will encourage having multiple Swift files as well.

# Summary

The Swift interpreter is a great way of learning how to program in Swift. It allows expressions, statements, and functions to be created and tested along with a command-line history that provides editing support. The basic collection types of arrays and collections, the standard data types, such as strings and numbers, optional values, and structures, were presented. Control flow and functions with positional, named, and variadic arguments, along with default values were also presented. Finally, the ability to write Swift scripts and run them from the command line was also demonstrated.

The next chapter will look at the other way of working with Swift code that is available on OS X, through the Xcode playground.

# 2
# Playing with Swift

Xcode ships with both a command-line interpreter (which was covered in *Chapter 1, Exploring Swift*) and a graphical interface called **playground** that can be used to prototype and test Swift code snippets. Code that is typed into the playground is compiled and executed interactively, which permits a fluid style of development. In addition, the user interface can present a graphical view of variables as well as a timeline, which can show how loops are executed. Finally, playgrounds can mix and match code and documentation leading to the possibility of providing example code as playgrounds and using playgrounds to learn how to use existing APIs and frameworks.

This chapter will present the following topics:

- How to create a playground
- Displaying values in the timeline
- Presenting objects with Quick Look
- Running asynchronous code
- Using playground's live documentation
- Creating multiple pages in a playground
- Limitations of playgrounds

# Getting started with playgrounds

When Xcode is started, the welcome screen is displayed with various options, including the ability to create a playground. The welcome screen can be shown with *Command + Shift + 1*, or by navigating to **Window | Welcome to Xcode**:

# Creating a playground

Using either the Xcode welcome screen (which can be opened by navigating to **Window | Welcome to Xcode**) or by navigating to **File | New | Playground**, create MyPlayground in a suitable location targeting **iOS**. Creating the playground on the Desktop will allow easy access to testing Swift code, but it can be located anywhere on the filesystem.

Playgrounds can be targeted either towards OS X applications or towards iOS applications. This can be configured when the playground is created, or by switching to the **Utilities** view by navigating to **View | Utilities | Show File Inspector** or pressing *Command + Option + 1* and changing the dropdown from OS X to iOS or vice versa:

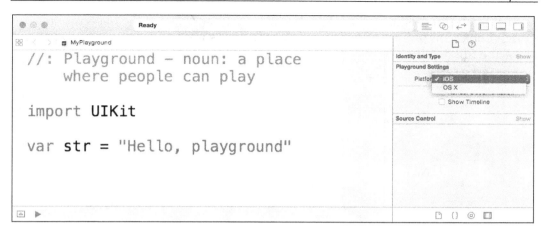

When initially created, the playground will have a code snippet that looks as follows:

```
// Playground - noun: a place where people can play
import UIKit
var str = "Hello, playground"
```

 Playgrounds targeting OS X will read `import Cocoa` instead.

On the right-hand side, a column will display the value of the code when each line is executed. By grabbing the vertical divider between the Swift code and the output, the output can be resized to show the full value of the text:

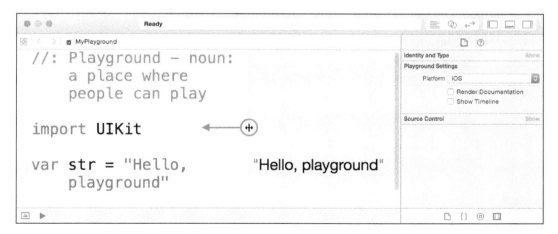

Alternatively, by moving the mouse over the right-hand side of the playground, the **Quick Look** icon (the eye symbol) will appear. If this is clicked, a pop-up box will show the full details:

# Viewing the console output

The console output can be viewed in the *Debug Area*. This can be shown by pressing *Command + Shift + Y* or by navigating to **View | Debug Area | Show Debug Area**. This will show the result of any `print` statements that are executed in the code.

Add a simple `for` loop to the playground:

```
for i in 1...12 {
  print("I is \(i)")
}
```

The output is shown in the debug area below:

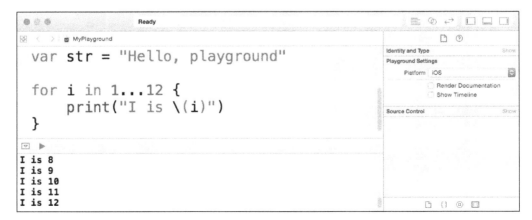

# Viewing the timeline

The timeline shows what the values were at a particular time. In the case of the print loop that was shown previously, the output was displayed in the **Debug Area**. However, it is possible to use the playground to inspect the value of an expression on a line without having to display it directly. In addition, results can be graphed to show how these values change over time. The value of the graph is shown in-line with the source code unlike previous versions of Xcode, which displayed them on the right.

Add another line above the `print` statement to calculate the result of executing an expression, `(i-6) * (i-7)`, and store it in a `j` constant.

On the line next to the variable definition, click on the add variable history (**+**) symbol, which is in the right-hand column (visible when the mouse moves over that area). After it is clicked , it will change to an (**o**) symbol and display the graph on the right-hand side. This can be done for the `print` statement as well:

```
for i in 1...12 {
    let j = (i-7) * (i-6)
    print("I is \(i)")
}
```

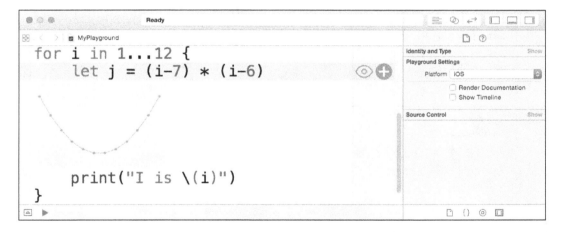

It is possible to display a timeline slider at the bottom of the window by selecting the **Show Timeline** checkbox in the **Utilities** area. This adds a timeline slider at the bottom, with a red tick mark, and this can be used to slide the vertical bar to see the exact value at certain points:

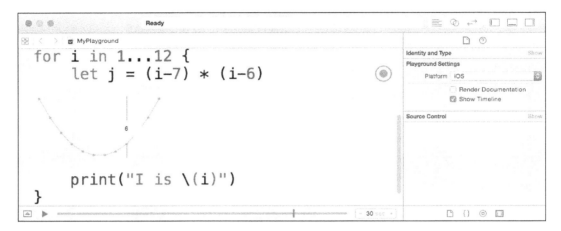

To display several values at once, use additional variables to hold the values and display them in the timeline as well:

```
for i in 1...12 {
    let j = (i-7) * (i-6)
    let k = i
    print("I is \(i)")
}
```

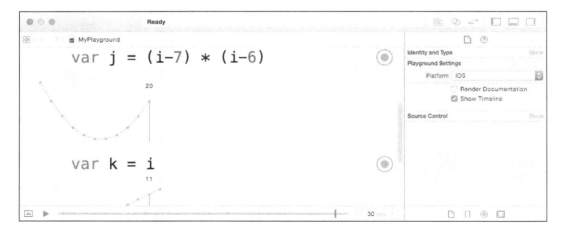

When the timeline slider is dragged, both values will be displayed at the same time.

# Displaying objects with Quick Look

The playground timeline can display objects as well as numbers and simple strings. It is possible to load and view images in a playground using classes, such as UIImage (or NSImage on OS X). These are known as *Quick Look supported objects*, and by default include:

- String (attributed and unattributed)
- Views
- Class and struct types (members are shown)
- Colors

 It is possible to build support for custom-based types into Swift by implementing a debugQuickLookObject method that returns a graphical view of the data.

# Showing colored labels

To show a colored label, a color needs to be obtained first. When building against iOS, this will be UIColor; but when building against OS X, it will be NSColor. The methods and types are largely equivalent between the two, but this chapter will demonstrate using iOS types.

A color can be acquired with an initializer or using one of the predefined colors that are exposed in Swift using methods, as follows:

```
import UIKit // AppKit for OS X
let blue = UIColor.blueColor() // NSColor.blueColor() for OS X
```

 In Xcode 7.1 and above, a color can be dragged in from a color picker into the Swift code directly where it will be translated as a color initializer with the specific hardcoded color values.

The color can be used as the textColor of a UILabel, which displays a text string in a particular size and color. The UILabel needs a size, which is represented by a CGRect, and this can be defined with an x and y position along with a width and height. The x and y positions are not relevant for playgrounds, and so, they can be left as zero:

```
let size = CGRect(x:0,y:0,width:200,height:100)
let label = UILabel(frame:size)// NSLabel for OS X
```

Finally, the text needs to be displayed in blue and with a larger font size:

```
label.text = str // from the first line of the code
label.textColor = blue
label.font = UIFont.systemFontOfSize(24) // NSFont for OS X
```

When the playground is run, the color and font are displayed in the timeline and are available for quick view. Even though the same `UILabel` instance is being shown, the timeline and the quick look values show a snapshot of the state of the object at each point, making it easy to see what has happened between changes:

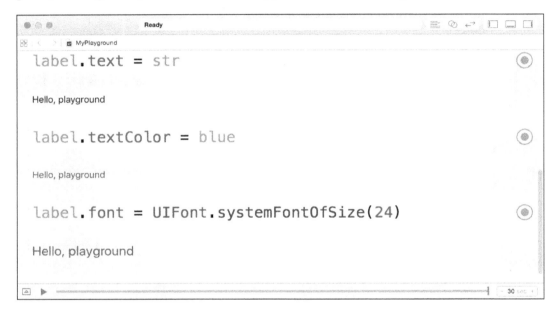

# Showing images

Images can be created and loaded into a playground using the `UIImage` constructor (or `NSImage` on OS X). Both take a `named` argument, which is used to find and load an image with the given name from the playground's `Resources` folder.

To copy a file into the playground's `Resources` folder, first download an image, such as `http://alblue.bandlem.com/images/AlexHeadshotLeft.png`, and save it as `alblue.png` in a suitable location such as the `Desktop`. In order to add it to the playground, the project navigator needs to be opened with *Command + 1* or by navigating to **View | Navigators | Show Project Navigator**. Once opened, the file can be dragged and dropped into the `Resources` element in the tree:

 Xcode 7.1 allows the image to be dragged directly into the source code area. It will populate a UIImage (or NSImage) as well as copy it to the resources area. Xcode 7.0 and below will just copy the full file path of the source if dragged in.

Alternatively, to download a logo with the command line, open Terminal.app and run the following commands:

```
$ mkdir MyPlayground.playground/Resources
```

```
$ curl http://alblue.bandlem.com/images/AlexHeadshotLeft.png >
MyPlayground.playground/Resources/alblue.png
```

An image can now be created in swift using the following:

```
let alblue = UIImage(named:"alblue")
```

 The location of the Resources that are associated with a playground can be seen in the **File Inspector** utilities view, which can be opened by pressing *Command + Option + 1*.

The created image can be displayed using **Quick Look** or by adding it to the value history:

 It is possible to use a URL to acquire an image by creating an NSURL with NSURL(string:"http://...")!, then loading the contents of the URL with NSData(contentsOfURL:)!, and finally, using UIImage(data:) to convert it to an image. However, as Swift will keep re-executing the code over and over again, the URL will be hit multiple times in a single debugging session without caching. It is recommended that NSData(contentsOfURL:) and similar networking classes be avoided in playgrounds.

# Advanced techniques

The playground has its own XCPlayground framework, which can be used to perform certain tasks. For example, individual values can be captured during loops for later analysis. It also permits asynchronous code to continue to execute once the playground has finished running.

# Capturing values explicitly

It is possible to explicitly add values to the timeline by importing the XCPlayground framework and using XCPlaygroundPage.currentPage, and calling captureValue with a value that should be displayed in the timeline. This takes an identifier, which is used both as the title and to group related data values in the same series. When the value history button is selected, it essentially inserts a call to captureValue with the value of the expression as the identifier.

For example, to add the logo to the timeline automatically:

```
import XCPlayground
let page = XCPlaygroundPage.currentPage
let alblue = UIImage(named:"alblue")
page.captureValue(alblue, withIdentifier:"Al Blue")
```

Opening the **Assistant Editor** will show the timeline along with the recorded values:

It is possible to use the identifier to group the data that is being shown in a loop, with the identifier representing categories of the values. For example, to display a list of all even and odd numbers between 1 and 6, the following code could be used:

```
for n in 1...6 {
  if n % 2 == 0 {
    page.captureValue(n,withIdentifier:"even")
    page.captureValue(0,withIdentifier:"odd")
  } else {
    page.captureValue(n,withIdentifier:"odd")
    page.captureValue(0,withIdentifier:"even")
  }
}
```

When executed, the result will look like:

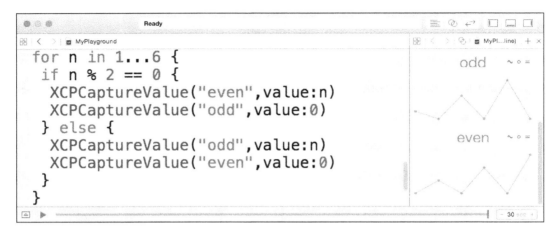

# Running asynchronous code

By default, when execution hits the end of the current playground page, the execution stops. In most cases this is desirable, but when asynchronous code is involved, execution may need to continue to run even if the main code has finished executing. This may be the case if networking data is involved or if there are multiple tasks whose results need to be synchronized.

For example, wrapping the previous even/odd split in an asynchronous call will result in no data being displayed:

```
dispatch_async(dispatch_get_main_queue()) {
  for n in 1...6 {
    // as before
  }
}
```

 This uses one of Swift's language features: the dispatch_async method which is actually a two-argument method that takes a queue and a block type. However, if the last argument is a block type, then it can be represented as a trailing closure rather than an argument.

To allow playground to continue executing after reaching the end, add the following assignment:

```
page.needsIndefiniteExecution = true
```

Although this suggests that the execution will run forever, it is limited to 30 seconds of runtime, or whatever the value is displayed at the bottom-right corner of the screen. This timeout can be changed by typing in a new value or using the **+** and **–** buttons to increase/decrease time by one second. In addition to this, the execution can be stopped by clicking the square icon on the lower left-hand side of the window:

# Playgrounds and documentation

Playgrounds can contain a mix of code and documentation. This allows a set of code, samples, and explanations to be mixed in with the playground itself.

# Learning with playgrounds

As playgrounds can contain a mixture of code and documentation, it makes them an ideal format to view annotated code snippets. In fact, Apple's Swift Tour book can be opened as a playground file.

Xcode documentation can be searched by navigating to **Help | Documentation and API Reference** or by pressing *Command + Shift + 0*. In the search dialog that is presented, type `Swift Tour` and then select the first result. The Swift Tour book should be presented in Xcode's help system, as follows:

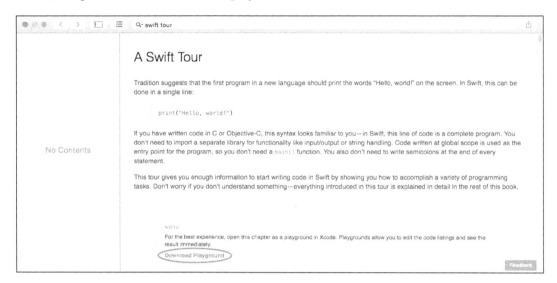

A link to download and open the documentation as a playground is given in the first section; if this is downloaded, it can be opened in Xcode as a standalone playground. This provides the same information, but it allows the code examples to be dynamic and show the results in the window:

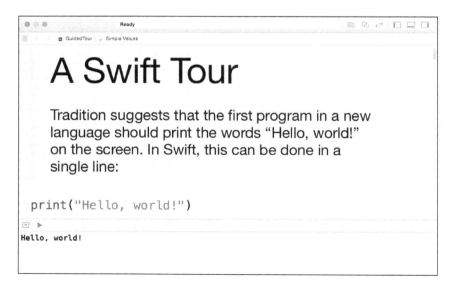

A key advantage of learning through playground-based documentation is that the code can be experimented with. In the *Simple Values* section of the documentation, where myVariable is assigned, the right-hand side of the playground shows the values. If the literal numbers are changed, the new values will be recalculated and shown on the right-hand side.

# Understanding the playground format

The playground is an *OS X bundle*, which means that it is a directory that looks like a single file. If the playground is selected either in TextEdit.app or in Finder, then it looks like a regular file:

Under the covers, it is actually a directory:

```
$ ls -F
MyPlayground.playground/
```

Inside the directory, there are a number of files:

```
$ ls -1 MyPlayground.playground/*
MyPlayground.playground/Contents.swift
MyPlayground.playground/Resources
MyPlayground.playground/contents.xcplayground
MyPlayground.playground/playground.xcworkspace
MyPlayground.playground/timeline.xctimeline
```

The files are as follows:

- The Contents.swift file, which is the Swift file that is created by default when a new playground is created, and this contains the code that is typed in for any new playground content

- The Resources directory, which was created earlier to hold the logo image

- The `contents.xcplayground` file, which is an XML table of contents of the files that make up the playground

- The `playground.xcworkspace`, which is used to hold metadata about the project in Xcode

- The `timeline.xctimeline`, which is the file containing timestamps of execution that are generated by the runtime when executing a Swift file and the timeline is open

The table of contents file defines which runtime environment is being targeted (for example, iOS or OS X) and a reference to the timeline file:

```
<playground version='5.0' target-platform='ios' requires-full-
environment='true' timelineScrubberEnabled='true' display-mode='raw'>
  <timeline fileName='timeline.xctimeline'/>
</playground>
```

An Xcode playground directory is deleted and recreated whenever changes are made in Xcode. Any `Terminal.app` windows that are open in that directory will no longer display any files. As a result, using external tools and editing the files in place may result in changes being lost. In addition, using ancient versions of control systems, such as SVN and CVS, may find their version control metadata being wiped out between saves. Xcode ships with the industry standard Git version control system, which should be preferred instead.

# Adding a page

By default, an Xcode playground has a single page open. However, for more comprehensive documentation examples, many separate pages may be preferable. For example, instead of creating a single very long page with subheadings (which may take a while to interpret and execute), additional pages can be added, each with their own specific examples. This also has the advantage of being able to interactively experiment with code as only the examples on a page need to be recalculated.

To add a new page to an existing playground, right-click on the **MyPlayground** top-level element in the project navigator and select the **New Playground Page** menu item. Alternatively, navigate to **File | New | Playground Page** or its keyboard shortcut, *Command + Option + N*. When this is done, the first page becomes **Untitled Page** and the newly added page becomes **Untitled Page 2**:

Pages can be reordered by dragging and dropping them in the project navigator on the left. They can also be renamed by selecting the page, then clicking it to reveal a text field that can be renamed. This is similar to renaming files in the `Finder`. The documentation's `@previous` and `@next` links allow the reader to navigate through the pages, as described in the following section.

When working with a playground with pages, the `contents.xcplayground` file's version number is updated to `6.0`, and a new `Pages` directory is created that sits alongside the `Resources` top-level folder. Inside the `Pages` directory, each page is represented as its own `.xcplaygroundpage` folder, which contains a `Contents.swift` file and an individual `timeline.xctimeline` file.

# Documenting code

Swift 2 adopts a new markup scheme for documentation, both for use with playgrounds but to also document Swift code. As a result, the documentation comments are described as applying to *Playground Comments* or *Symbol Documentation*.

Playground comments start with `//:` for single-line comments, and uses `/*:` and `*/` for block-level comments. These are rendered in playgrounds as in-line documentation, and they replace the nested HTML that existed in prior versions of Xcode. Markup defaults to showing as raw text, but the rendered content can be seen by navigating to **Editor | Show Rendered Markup**. To toggle it back to display the raw markup and allow the text to be edited, navigate to **Editor | Show Raw Markup**. This setting is also persisted in the `xcplayground` file with the `display-mode='rendered'` or `display-mode='raw'` attribute.

Symbol documentation starts with /// for single-line comments, and uses /** and */ for block-level comments. Symbol documentation applies to variables and constants, functions, and types. Only one type of symbol documentation comment (either single-line or multiline, but not both) may be present above a symbol definition. Multiple contiguous single-line comments will be merged into a single block.

 Symbol documentation can be revealed by pressing *Command + Control + ?* while the cursor is over an identifier, or by pressing *Alt* and clicking on the identifier in Xcode.

Both playground and symbol documentation allow some markup to be used for text formatting purposes. In addition, there are certain symbol *format commands* that can be specified with a hyphen, followed by the command name, and then a colon. These are used to introduce documentation, for example, for a specific parameter of a function, the return type, or what errors are thrown.

# Playground navigation documentation

It is possible to create navigational links between pages in a multipage playground. Each page has a name (which starts off as **Untitled Page**, **Untitled Page 2**, and so on) but can be renamed in the project navigator.

To rename a page, open the project navigator with *Command + 1* and then select the page in the navigator view. The name can be made editable by double-clicking on the page name, which turns it into a text field:

Links to specific pages are performed with a link, which is represented as
`[Link Name](Page%20Name)`. For example, to create a link to the first page
that was just shown, the following can be used:

```
//: Go back to the [first page](Page%20One)
```

 Spaces in the page name need to be URL escaped, so a
space is represented as `%20`. Using + does not work.

As page names may be fragile, it is recommended to use the **Next** and **Previous** links
instead. These can be represented using the `@next` and `@previous` special identifiers
as the page names, as follows:

```
//: Go back to [the previous page](@previous),
//: or move forward to [the next page](@next).
```

Using `@next` and `@previous` is recommended in order to chain multiple pages
together because it allows pages to be reordered without requiring any changes
to the content. Pages can be reordered in the project navigator by dragging and
dropping projects up and down the order.

 Page navigation is only available in playgrounds.

# Text formatting

The playground and symbol documentation can use a number of different
formatting styles using a markup language to represent different types of text. These
include the following:

- Bulleted lists, which use one of the *, +, or – characters as the bullet, followed
  by a single space, and the text
- Numbered lists, which use a number, followed by a period, a single space,
  and the text
- Horizontal rules, which use four dashes ---- to generate a horizontal rule in
  the text
- Block quotations, which start each line with > followed by a single space

- Block code, which is either indented four spaces in from the start, or begin and end with ````

- Headings, using #, ##, or ### for level 1, 2, or 3 headings, respectively. Alternatively, heading level 1 can use a === underneath the title and heading level 2 can use --- underneath the title

 Exactly a single space is required between the end of the list delimiter (the bullet or the period) and the following text; otherwise it will not be rendered as expected

In addition to the block-level formatting, it is possible to use in-line formatting elements:

- Code can be represented with `backticks` around the words
- Text can be emphasized in italics _like this_ or *like this*
- Text can be marked as bold using __this__ or **this**

Images and links can also be used in documentation:

- Images are represented with ![Alternate Text](url "hover text")
- Links are represented with [link text](url)
- Links can also be declared with [link title]: url "hover text", and then referred to later with [link title]

For example, here is a markup block consisting of many single-line comments, which will be concatenated into a single documentation block:

```
//: # Example Documentation
//: Navigate to the [previous](@previous) or [next](@next) page
//: ----
//: Numbered lists:
//: 1. First item
//: 2. Second item
//: 3. Third item
//:
//: Bulleted lists:
//: * First item
//: * Second item
//:    + child item
//:    + child item
//: * Third item
//:
//: How to do loops in Swift using `for`:
```

```
//:
//:      for i in 1...12 {
//:        print("Looping \(i)")
//:      }
//:
//: > This is a block quote
//: > which is merged together
//: > using _italics_ or **bold**
//:
//: Link to [AlBlue's Blog](http://alblue.bandlem.com)
//: Image of ![AlBlue](http://alblue.bandlem.com/images/
AlexHeadshotLeft.png "AlBlue")
```

When viewed in a rendered markup view, it will look like:

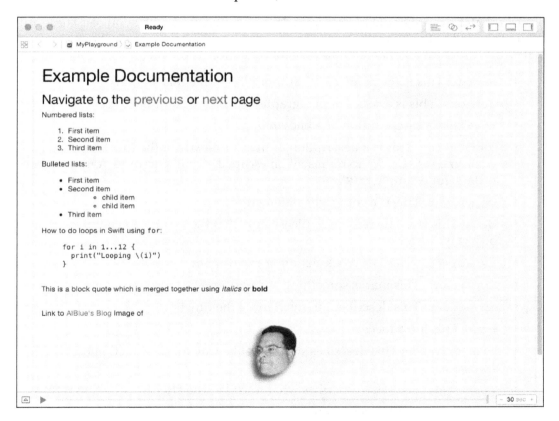

# Symbol documentation

When writing documentation for symbols (variables, constants, functions, and so on), additional commands can be used to indicate particular values. These are represented with a – dash, followed by a command name, and then a colon. These include the following:

- `- parameter name: description`: This is the `description` of a `parameter` called `name` in the function or method
- `- parameters:`: This is a group of related `parameter` elements
- `- returns: description`: This is the `description` of the return result
- `- throws: description`: This is the `description` of any errors that may occur

There are also a number of field commands that can be used with symbols. These are all represented with – `name: description`, and they all have exactly the same effect but with a different text name in the symbol reference. These include the following:

- `author`: This is the name of the author who wrote this section.
- `authors`: This is a series of paragraphs with one author name per paragraph.
- `bug`: This is a description of a known bug.
- `complexity`: This is a description of the complexity of the function, such as `O(1)` or `O(N)`. Use `\*` to represent an escaped `*` character or to denote higher orders, for example, `O(N\*N)`.
- `copyright`: This is a copyright statement.
- `date`: This is a date reference; please note that the text field is not parsed in any way.
- `experiment`: This is a block-denoting experiment.
- `important`: This marks something as important.
- `invariant`: This describes an invariant of the function.
- `note`: This introduces a note.
- `precondition`: This describes what must be true for the function that has to be called.
- `postcondition`: This describes what must be true after the function returns.
- `remark`: This adds general notes to the symbol.
- `requires`: This notes what is required, such as module dependencies, or a minimum version of the operating system.
- `seealso`: This adds a **See Also** link to the documentation.

- since: This is documentation indicating when the functionality first arrived.

- todo: This adds a note to do later.

- version: This documents the version number of the location.

- warning: This adds a warning note.

The following combines a number of these documentation examples into a multi-line documentation block:

```
/**
Returns the string in SHOUTY CAPS
- parameter input: the input string
- author: Alex Blewitt
- returns: The input string, but in upper case
- throws: No errors thrown
- note: Please don't shout
- seealso: String.uppercaseString
- since: 2015
 */
func shout(input:String) -> String {
  return input.uppercaseString
}
```

When the mouse hovers over the shout function, the following documentation will be seen:

| Declaration | func shout(input: String) -> String |
|---|---|
| Description | Returns the string in SHOUTY CAPS |
| | **Author:**<br>Alex Blewitt |
| | **Note:**<br>Please don't shout |
| | **See also:**<br>String.uppercaseString |
| | **Since:**<br>2015 |
| Parameters | input   the input string |
| Throws | No errors thrown |
| Returns | The input string, but in upper case |
| Declared In | Example Documentation.xcplaygroundpage |

# Limitations of playgrounds

Although playgrounds can be very powerful for interacting with code, there are some limitations that are worth being aware of. There is no debugging support in the playground, so it is not possible to add a breakpoint and use the debugger and find out what are the values.

Given that the UI allows tracking values—and that it's very easy to add new lines with just the value to be tracked—this is not much of a hardship. Other limitations of playgrounds include the following:

- Only the simulator can be used for the execution of iOS-based playgrounds. This prevents the use of hardware-specific features that may only be present on a device.

- The performance of playground scripts is mainly based on how many lines are executed and how much output is saved by the debugger. It should not be used to test the performance of performance-sensitive code.

- Although the playground is well suited to present user interface components, it cannot be used for user input.

- Anything requiring entitlements (such as in-app purchases or access to iCloud) is not possible in playground at the current time of writing.

# Summary

This chapter presented playgrounds, an innovative way of running Swift code with graphical representation of values and introspection of running code. Both expressions and timeline were presented as a way of showing the state of the program at any time, as well as graphically inspecting objects using Quick Look. The XCPlayground framework can also be used to record specific values and allow asynchronous code to be executed.

The next chapter will look at how to create an iOS application with Swift.

# 3

# Creating an iOS Swift App

After the release of Xcode 6 in 2014, it has been possible to build Swift applications for iOS and OS X and submit them to the App Store for publication. This chapter will present both a single view application and a master-detail application, and use these to explain the concepts behind iOS applications, as well as introduce classes in Swift.

This chapter will present the following topics:

- How iOS applications are structured
- Single-view iOS applications
- Creating classes in Swift
- Protocols and enums in Swift
- Using XCTest to test Swift code
- Master-detail iOS applications
- The AppDelegate and ViewController classes

## Understanding iOS applications

An iOS application is a compiled executable along with a set of supporting files in a bundle. The application bundle is packaged into an archive file to be installed onto a device or upload to the App Store.

 Xcode can be used to run iOS applications in a simulator, as well as testing them on a local device. Submitting an application to the App Store requires a developer signing key, which is included as part of the Apple Developer Program at https://developer.apple.com.

Most iOS applications to date have been written in Objective-C, a crossover between C and Smalltalk. With the advent of Swift, it is likely that many developers will move at least parts of their applications to Swift for performance and maintenance reasons.

Although Objective-C is likely to be around for a while, it is clear that Swift is the future of iOS development and probably OS X as well. Applications contain a number of different types of files, which are used both at compile time and also at runtime. These files include the following:

- The `Info.plist` file, which contains information about which languages the application is localized for, what the identity of the application is, and the configuration requirements, such as the supported interface types (iPad, iPhone, and Universal), and orientations (Portrait, Upside Down, Landscape Left, and Landscape Right)

- Zero or more *interface builder* files with a `.xib` extension, which contain user interface screens (which supersedes the previous `.nib` files)

- Zero or more *image asset* files with a `.xcassets` extension, which store groups of related icons at different sizes, such as the application icon or graphics for display on screen (which supersedes the previous `.icns` files)

- Zero or more *storyboard* files with a `.storyboard` extension, which are used to coordinate between different screens in an application

- One or more `.swift` files that contain application code

# Creating a single-view iOS application

A single-view iOS application is one where the application is presented in a single screen, without any transitions or other views. This section will show how to create an application that uses a single view without storyboards. (Storyboards are covered in *Chapter 4, Storyboard Applications with Swift and iOS*.)

When Xcode starts, it displays a welcome message that includes the ability to create a new project. This welcome message can be redisplayed at any time by navigating to **Window | Welcome to Xcode** or by pressing *Command + Shift + 1*.

Using the welcome dialog's **Create a new Xcode project** option, or navigating to **File | New | Project...**, or by pressing *Command + Shift + N*, create a new **iOS** project with **Single View Application** as the template, as shown in the following screenshot:

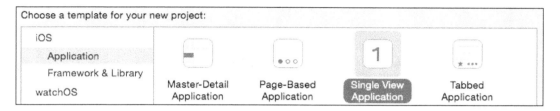

When the **Next** button is pressed, the new project dialog will ask for more details. The product name here is SingleView with appropriate values for **Organization Name** and **Identifier**. Ensure that the language selected is **Swift** and the device type is **Universal**:

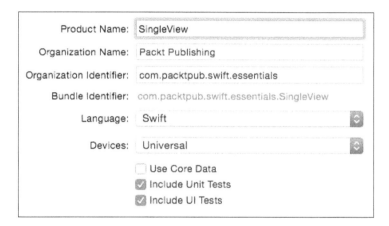

 The **Organization Identifier** is a reverse domain name representation of the organization, and the **Bundle Identifier** is the concatenation of the **Organization Identifier** with the **Product Name**. Publishing to the App Store requires that the **Organization Identifier** be owned by the publisher and is managed in the online developer center at https://developer.apple.com/membercenter/.

When **Next** is pressed, Xcode will ask where to save the project and whether a repository should be created. The selected location will be used to create the product directory, and an option to create a Git repository will be offered.

In 2014, Git became the most widely used version control system, surpassing all other distributed and centralized version control systems. It would be foolish not to create a Git repository when creating a new Xcode project.

When **Create** is pressed, Xcode will create the project, set up template files, and then initialize the Git repository locally or on a shared server.

Press the triangular play button at the top-left of Xcode to launch the simulator:

If everything has been set up correctly, the simulator will start with a white screen and the time and battery shown at the top of the screen:

# Removing the storyboard

The default template for a single-view application includes a *storyboard*. This creates the view for the first (only) screen and performs some additional setup behind the scenes. To understand what happens, the storyboard will be removed and replaced with code instead.

 Most applications are built with one or more storyboards. It is being removed here for demonstration purposes only; refer to the *Chapter 4, Storyboard Applications with Swift and iOS*, for more information on how to use storyboards.

The storyboard can be deleted by going to the project navigator, finding the `Main.storyboard` file, and pressing the *Delete* key or selecting **Delete** from the context-sensitive menu. When the confirmation dialog is shown, select the **Move to Trash** option to ensure that the file is deleted rather than just being removed from the list of files that Xcode knows about.

To see the project navigator, press *Command + 1* or navigate to
**View | Navigators | Show Project Navigator**.

Once the `Main.storyboard` file has been deleted, it needs to be removed from
`Info.plist`, to prevent iOS from trying to open it at startup. Open the `Info.plist`
file under the `Supporting Files` folder of `SingleView`. A set of key-value pairs will
be displayed; clicking on the **Main storyboard file base name** row will present the
**(+)** and **(-)** options. Clicking on the delete icon **(-)** will remove the line:

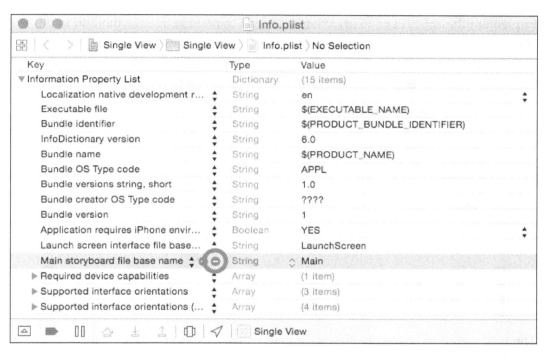

Now when the application is started, a black screen will be displayed.

 There are multiple `Info.plist` files that are created by Xcode's
template; one file is used for the real application, while the other files are
used for the test applications that get built when running tests. Testing is
covered in the *Subclasses and testing in Swift* section later in this chapter.

# Setting up the view controller

The *view controller* is responsible for setting up the view when it is activated. Typically, this is done through either the storyboard or the interface file. As these have been removed, the window and the view controller need to be instantiated manually.

When iOS applications start, `application:didFinishLaunchingWithOptions:` is called on the corresponding `UIApplicationDelegate`. The optional `window` variable is initialized automatically when it is loaded from an interface file or a storyboard, but it needs to be explicitly initialized if the user interface is being implemented in code.

Implement the `application:didFinishLaunchingWithOptions:` method in the `AppDelegate` class as follows:

```
@UIApplicationMain
class AppDelegate: UIResponder, UIApplicationDelegate {
  var window: UIWindow?
  func application(application: UIApplication,
   didFinishLaunchingWithOptions launchOptions:
   [NSObject:AnyObject]?) -> Bool {
   window = UIWindow()
   window?.rootViewController = ViewController()
   window?.makeKeyAndVisible()
   return true
  }
}
```

 To open a class by name, press *Command + Shift + O* and type in the class name. Alternatively, navigate to **File | Open Quickly...**

The final step is to create the view's content, which is typically done in the `viewDidLoad` method of the `ViewController` class. As an example user interface, a `UILabel` will be created and added to the view. Each view controller has an associated `view` property, and child views can be added with the `addSubview` method. To make the view stand out, the background of the view will be changed to black and the text color will be changed to white:

```
class ViewController: UIViewController {
  override func viewDidLoad() {
   super.viewDidLoad()
   view.backgroundColor = UIColor.blackColor()
   let label = UILabel(frame:view.bounds)
   label.textColor = UIColor.whiteColor()
```

```
    label.textAlignment = .Center
    label.text = "Welcome to Swift"

    view.addSubview(label)
  }
}
```

This creates a label, which is sized to the full size of the screen, with a white text color and a centered text alignment. When run, this displays **Welcome to Swift** on the screen.

Typically, views will be implemented in their own class rather than being in-lined into the view controller. This allows the views to be reused in other controllers. This technique will be demonstrated in the next chapter.

When the screen is rotated, the label will be rotated off screen. Logic would need to be added in a real application to handle rotation changes in the view controller, such as `willRotateToInterfaceOrientation`, and to appropriately add rotations to the views using the `transform` property of the view. Usually, an interface builder file or storyboard would be used so that this is handled automatically.

# Swift classes, protocols, and enums

Almost all Swift applications will be object oriented. *Chapter 1, Exploring Swift*, and *Chapter 2, Playing with Swift*, both demonstrated functional and procedural Swift code. Classes, such as `Process` from the `CoreFoundation` framework, and `UIColor` and `UIImage` from the `UIKit` framework, were used to demonstrate how classes can be used in applications. This section describes how to create classes, protocols, and enums in Swift.

## Classes in Swift

A class is created in Swift using the `class` keyword, and braces are used to enclose the class body. The body can contain variables called *properties*, as well as functions called *methods*, which are collectively referred to as *members*. Instance members are unique to each instance, while static members are shared between all instances of that class.

Classes are typically defined in a file named for the class; so a `GitHubRepository` class would typically be defined in a `GitHubRepository.swift` file. A new Swift file can be created by navigating to **File | New | File...** and selecting the **Swift File** option under **iOS**. Ensure that it is added to the **Tests** and **UITests** targets as well. Once created, implement the class as follows:

```
class GitHubRepository {
  var id:UInt64 = 0
  var name:String = ""
  func detailsURL() -> String {
    return "https://api.github.com/repositories/\(id)"
  }
}
```

This class can be instantiated and used as follows:

```
let repo = GitHubRepository()
repo.id = 1
repo.name = "Grit"
repo.detailsURL() // returns https://api.github.com/repositories/1
```

It is possible to create static members, which are the same for all instances of a class. In the `GitHubRepository` class, the `api` URL is likely to remain the same for all invocations, so it can be refactored into a `static` property:

```
class GitHubRepository {
  // does not work in Swift 1.0 or 1.1
  static let api = "https://api.github.com"
  ...
  class func detailsURL(id:String) -> String {
    return "\(api)/repositories/\(id)"
  }
}
```

Now, if the `api` URL needs to be changed (for example, to support mock testing or to support an in-house GitHub Enterprise server), there is a single place to change it. Before Swift 2, a **class variables are not yet supported** error message may be displayed.

To use static variables in Swift prior to version 2, a different approach must be used. It is possible to define *computed properties*, which are not stored but are calculated on demand. These have a *getter* (also known as an *accessor*) and optionally a *setter* (also known as a *mutator*). The previous example can be rewritten as follows:

```
class GitHubRepository {
  class var api:String {
    get {
      return "https://api.github.com"
```

```
    }
  }
  func detailsURL() -> String {
    return "\(GitHubRepository.api)/repositories/\(id)"
  }
}
```

Although this is logically a read-only constant (there is no associated `set` block), it is not possible to define the `let` constants with accessors.

To refer to a class variable, use the type name—which in this case is `GitHubRepository`. When the `GitHubRepository.api` expression is evaluated, the body of the getter is called.

# Subclasses and testing in Swift

A simple Swift class with no explicit parent is known as a *base class*. However, classes in Swift frequently *inherit* from another class by specifying a superclass after the class name. The syntax for this is `class SubClass:SuperClass{...}`.

Tests in Swift are written using the **XCTest** framework, which is included by default in Xcode templates. This allows an application to have tests written and then executed in place to confirm that no bugs have been introduced.

 XCTest replaces the previous testing framework OCUnit.

The `XCTest` framework has a base class called `XCTestCase` that all tests inherit from. Methods beginning with `test` (and that take no arguments) in the test case class are invoked automatically when the tests are run. Test code can indicate success or failure by calling the `XCTAssert*` functions, such as `XCTAssertEquals` and `XCTAssertGreaterThan`.

Tests for the `GitHubRepository` class conventionally exist in a corresponding `GitHubRepositoryTest` class, which will be a subclass of `XCTestCase`. Create a new Swift file by navigating to **File | New | File...** and choosing a **Swift File** under the **Source** category for **iOS**. Ensure that the **Tests** and **UITests** targets are selected but the application target is not. It can be implemented as follows:

```
import XCTest
class GitHubRepositoryTest: XCTestCase {
  func testRepository() {
    let repo = GitHubRepository()
    repo.id = 1
```

```
      repo.name = "Grit"
      XCTAssertEqual(
        repo.detailsURL(),
        "https://api.github.com/repositories/1",
        "Repository details"
      )
    }
  }
```

Make sure that the `GitHubRepositoryTest` class is added to the test targets. If not added when the file is created, it can be done by selecting the file and pressing *Command + Option + 1* to show the **File Inspector**. The checkbox next to the test target should be selected. Tests should never be added to the main target. The `GitHubRepository` class should be added to both test targets:

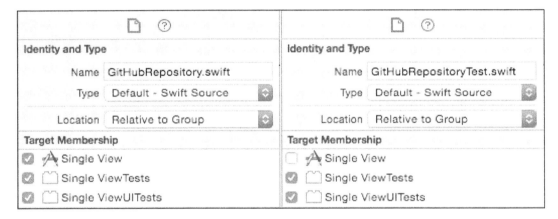

When the tests are run by pressing *Command + U* or by navigating to **Product | Test**, the results of the test will be displayed. Changing either the implementation or the expected test result will demonstrate whether the test is being executed correctly.

Always check whether a failing test causes the build to fail; this will confirm that the test is actually being run. For example, in the `GitHubRepositoryTest` class, modify the URL to remove `https` from the front and check whether a test failure is shown. There is nothing more useless than a correctly implemented test that never runs.

# Protocols in Swift

A **protocol** is similar to an interface in other languages; it is a named type that has method signatures but no method implementations. Classes can implement zero or more protocols; when they do, they are said to **adopt** or **conform** to the protocol. A protocol may have a number of methods that are either **required** (the default) or **optional** (marked with the `optional` keyword).

 Optional protocol methods are only supported when the protocol is marked with the `@objc` attribute. This declares that the class will be backed by an `NSObject` class for interoperability with Objective-C. Pure Swift protocols cannot have optional methods.

The syntax to define a protocol looks similar to the following:

```
protocol GitHubDetails {
  func detailsURL() -> String
  // protocol needs @objc if using optional protocols
  // optional doNotNeedToImplement()
}
```

 Protocols cannot have functions with default arguments. Protocols can be used with the `struct`, `class`, and `enum` types unless the `@objc` class attribute is used; in which case, they can only be used against Objective-C classes or enums.

Classes conform to protocols by listing the protocol names after the class name, similar to a superclass.

 When a class has both a superclass and one or more protocols, the superclass must be listed first.

```
class GitHubRepository: GitHubDetails {
  func detailsURL() -> String {
    // implementation as before
  }
}
```

The `GitHubDetails` protocol can be used as a type in the same places as an existing Swift type, such as a variable type, method return type, or argument type.

Protocols are widely used in Swift to allow callbacks from frameworks that would, otherwise, not know about specific callback handlers. If a superclass was required instead, then a single class cannot be used to implement multiple callbacks. Common protocols include `UIApplicationDelegate`, `Printable`, and `Comparable`.

# Enums in Swift

The final concept to understand in Swift is *enumeration*, or *enum* for short. An enum is a closed set of values, such as `North`, `East`, `South`, and `West`, or `Up`, and `Down`.

An enumeration is defined using the `enum` keyword, followed by a type name, and a block, which contains the `case` keywords followed by comma-separated values as follows:

```
enum Suit {
  case Clubs, Diamonds, Hearts // many on one line
  case Spades // or each on separate lines
}
```

Unlike C, enumerated values do not have a specific type by default, so they cannot generally be converted to and from an integer value. Enumerations can be defined with **raw values** that allow conversion to and from integer values. Enum values are assigned to variables using the type name and the `enum` name:

```
var suit:Suit = Suit.Clubs
```

However, if the type of the expression is known, then the type prefix does not need to be explicitly specified; the following form is much more common in Swift code:

```
var suit:Suit = .Clubs
```

# Raw values

For the `enum` values that have specific meanings, it is possible to extend the `enum` from a different type, such as `Int`. These are known as *raw values*:

```
enum Rank: Int {
  case Two = 2, Three, Four, Five, Six, Seven, Eight, Nine, Ten
  case Jack, Queen, King, Ace
}
```

A raw value enum can be converted to and from its raw value with the `rawValue` property and the failable initializer `Rank(rawValue:)` as follows:

```
Rank.Two.rawValue == 2
Rank(rawValue:14)! == .Ace
```

 The failable initializer returns an optional enum value, because the equivalent `Rank` may not exist. The expression `Rank(rawValue:0)` will return `nil`, for example.

## Associated values

Enums can also have *associated values*, such as a value or case class in other languages. For example, a combination of a `Suit` and a `Rank` can be combined to form a `Card`:

```
enum Card {
  case Face(Rank, Suit)
  case Joker
}
```

Instances can be created by passing values into an `enum` initializer:

```
var aceOfSpades: Card = .Face(.Ace,.Spades)
var twoOfHearts: Card = .Face(.Two,.Hearts)
var theJoker: Card = .Joker
```

The associated values of an `enum` instance cannot be extracted (as they can with properties of a `struct`), but the `enum` value can be accessed by pattern matching in a `switch` statement:

```
var card = aceOfSpades // or theJoker or twoOfHearts ...
switch card {
  case .Face(let rank, let suit):
    print("Got a face card \(rank) of \(suit)");
  case .Joker:
    print("Got the joker card")
}
```

The Swift compiler will require that the `switch` statement be exhaustive. As the `enum` only contains these two types, no `default` block is needed. If another `enum` value is added to `Card` in the future, the compiler will report an error in this `switch` statement.

# Creating a master-detail iOS application

Having covered how classes, protocols, and enums are defined in Swift, a more complex master-detail application can be created. A master-detail application is a specific type of iOS application that initially presents a master table view, and when an individual element is selected, a secondary details view will show more information about the selected item.

Using the **Create a new Xcode project** option from the welcome screen, or by navigating to **File | New | Project...** or by pressing *Command + Shift + N*, create a new project and select **Master-Detail Application** from the **iOS Application** category:

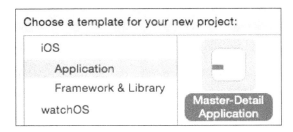

In the subsequent dialog, enter appropriate values for the project, such as the name (`MasterDetail`), the organization identifier (typically based on the reverse DNS name), ensure that the **Language** dropdown reads **Swift** and that it is targeted for **Universal** devices:

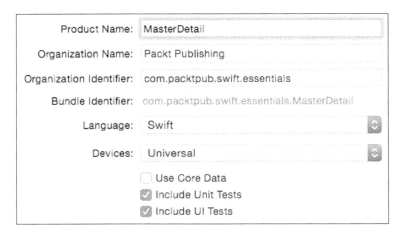

When the project is created, an Xcode window will open containing all the files that are created by the wizard itself, including the `MasterDetail.app` and `MasterDetailTests.xctest` products. The `MasterDetail.app` is a bundle that is executed by the simulator or a connected device, while the `MasterDetailTests.xctest` and `MasterDetailsUITests.xctest` products are used to execute unit tests for the application's code.

The application can be launched by pressing the triangular play button on the top-left corner of Xcode or by pressing *Command + R*, which will run the application against the currently selected target.

After a brief compile and build cycle, the iOS Simulator will open with a master page that contains an empty table, as shown in the following screenshot:

The default `MasterDetail` application can be used to add items to the list by clicking on the add (**+**) button on the top-right corner of the screen. This will add a new timestamped entry to the list.

When this item is clicked, the screen will switch to the details view, which, in this case, presents the time in the center of the screen:

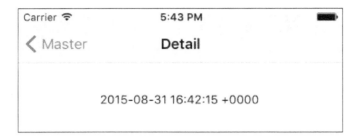

This kind of master-detail application is common in iOS applications for displaying a top-level list (such as a shopping list, a set of contacts, to-do notes, and so on) while allowing the user to tap to see the details.

There are three main classes in the master-detail application:

- The `AppDelegate` class is defined in the `AppDelegate.swift` file, and it is responsible for starting the application and set up the initial state
- The `MasterViewController` class is defined in the `MasterViewController.swift` file, and it is used to manage the first (master) screen's content and interactions
- The `DetailViewController` class is defined in the `DetailViewController.swift` file, and it is used to manage the second (detail) screen's content

In order to understand what the classes do in more detail, the next three sections will present each of them in turn.

> The code that is generated in this section was created from Xcode 7.0, so the templates might differ slightly if using a different version of Xcode. An exact copy of the corresponding code can be acquired from the Packt website or from this book's GitHub repository at `https://github.com/alblue/com.packtpub.swift.essentials/`.

# The AppDelegate class

The `AppDelegate` class is the main entry point to the application. When a set of Swift source files are compiled, if the `main.swift` file exists, it is used as the entry point for the application by running that code. However, to simplify setting up an application for iOS, a `@UIApplicationMain` special attribute exists that will both synthesize the `main` method and set up the associated class as the application delegate.

The `AppDelegate` class for iOS extends the `UIResponder` class, which is the parent of all the UI content on iOS. It also adopts two protocols, `UIApplicationDelegate`, and `UISplitViewControllerDelegate`, which are used to provide callbacks when certain events occur:

```
@UIApplicationMain
class AppDelegate: UIResponder, UIApplicationDelegate,
  UISplitViewControllerDelegate {
  var window: UIWindow?

  . . .

}
```

 On OS X, the `AppDelegate` class will be a subclass of `NSApplication` and will adopt the `NSApplicationDelegate` protocol.

The synthesized `main` function calls the `UIApplicationMain` method that reads the `Info.plist` file. If the `UILaunchStoryboardName` key exists and points to a suitable file (the `LaunchScreen.xib` interface file in this case), it will be shown as a splash screen before doing any further work. After the rest of the application has loaded, if the `UIMainStoryboardFile` key exists and points to a suitable file (the `Main.storyboard` file in this case), the storyboard is launched and the initial view controller is shown.

The storyboard has references to the `MasterViewController` and `DetailViewController` classes. The `window` variable is assigned to the storyboard's window.

The `application:didFinishLaunchingWithOptions` is called once the application has started. It is passed with a reference to the `UIApplication` instance and a dictionary of options that notifies how the application has been started:

```
func application(
 application: UIApplication,
 didFinishLaunchingWithOptions launchOptions:
  [NSObject: AnyObject]?) -> Bool {
 // Override point for customization after application launch.
  . . .
}
```

In the sample `MasterDetail` application, the `application:didFinishLaunching WithOptions` method acquires a reference to the `splitViewController` from the explicitly unwrapped optional `window`, and the `AppDelegate` is set as its delegate:

```
let splitViewController =
  self.window!.rootViewController as! UISplitViewController
splitViewController.delegate = self
```

 The `... as! UISplitViewController` syntax performs a type cast so that the generic `rootViewController` can be assigned to the more specific type; in this case, `UISplitViewController`. An alternative version `as?` provides a runtime checked cast, and it returns an optional value that either contains the value with the correctly casted type or `nil` otherwise. The difference with `as!` is a runtime error will occur if the item is not of the correct type.

Finally, a `navigationController` is acquired from the `splitViewController`, which stores an array of `viewControllers`. This allows the `DetailView` to display a button on the left-hand side to expand the details view if necessary:

```
let navigationController = splitViewController.viewController
  [splitViewController.viewControllers.count-1]
  as! UINavigationController
navigationController.topViewController
  .navigationItem.leftBarButtonItem =
  splitViewController.displayModeButtonItem()
```

The only difference this makes is when running on a wide-screen device, such as an iPhone 6 Plus or an iPad, where the views are displayed side-by-side in landscape mode. This is a new feature in iOS 8 applications.

Otherwise, when the device is in portrait mode, it will be rendered as a standard back button:

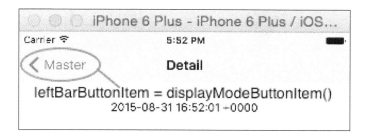

The method concludes with `return true` to let the OS know that the application has opened successfully.

# The MasterViewController class

The `MasterViewController` class is responsible for coordinating the data that is shown on the first screen (when the device is in portrait orientation) or the left-half of the screen (when a large device is in landscape orientation). This is rendered with a `UITableView`, and data is coordinated through the parent `UITableViewController` class:

```
class MasterViewController: UITableViewController {
  var detailViewcontroller: DetailViewController? = nil
  var objects = [AnyObject]()
  override func viewDidLoad() {…}
  func insertNewObject(sender: AnyObject) {…}
  …
}
```

The `viewDidLoad` method is used to set up or initialize the view after it has loaded. In this case, a `UIBarButtonItem` is created so that the user can add new entries to the table. The `UIBarButtonItem` takes a `@selector` in Objective-C, and in Swift is treated as a string literal convertible (so that `"insertNewObject:"` will result in a call to the `insertNewObject` method). Once created, the button is added to the navigation on the right-hand side, using the standard `.Add` type which will be rendered as a **+** sign on the screen:

```
override func viewDidLoad() {
  super.viewDidLoad()
  self.navigationItem.leftBarButtonItem = self.editButtonItem()
  let addButton = UIBarButtonItem(
    barButtonSystemItem: .Add, target: self,
```

```
    action: "insertNewObject:")
  self.navigationItem.rightBarButtonItem = addButton
  if let split = self.splitViewController {
    let controllers = split.viewControllers
    self.detailViewController = (controllers[controllers.count-1] as!
UINavigationController).topViewController as? DetailViewController
  }
```

The objects are NSDate values, and are stored inside the class as an Array of AnyObject elements. The insertNewObject method is called when the + button is pressed, and it creates a new NSDate instance which is then inserted into the array. The sender event is passed as an argument of the AnyObject type, which will be a reference to the UIBarButtonItem (although it is not needed or used here):

```
func insertNewObject(sender: AnyObject) {
  objects.insertObject(NSDate.date(), atIndex: 0)
  let indexPath = NSIndexPath(forRow: 0, inSection: 0)
  self.tableView.insertRowsAtIndexPaths(
    [indexPath], withRowAnimation: .Automatic)
}
```

The UIBarButtonItem class was created before blocks were available on iOS devices, so it uses the older Objective-C @selector mechanism. A future release of iOS may provide an alternative that takes a block, in which case Swift functions can be passed instead.

The parent class contains a reference to the tableView, which is automatically created by the storyboard. When an item is inserted, the tableView is notified that a new object is available. Standard UITableViewController methods are used to access the data from the array:

```
override func numberOfSectionsInTableView(
 tableView: UITableView) -> Int {
  return 1
}
override func tableView(tableView: UITableView,
 numberOfRowsInSection section: Int) -> Int {
  return objects.count
}
override func tableView(tableView: UITableView,
 cellForRowAtIndexPath indexPath: NSIndexPath) -> UITableViewCell{
  let cell = tableView.dequeueReusableCellWithIdentifier(
    "Cell", forIndexPath: indexPath)
  let object = objects[indexPath.row] as! NSDate
```

```
    cell.textLabel!.text = object.description
    return cell
}
override func tableView(tableView: UITableView,
  canEditRowAtIndexPath indexPath: NSIndexPath) -> Bool {
    return true
}
```

The `numberOfSectionsInTableView` function returns 1 in this case, but a `tableView` can have multiple sections; for example, to permit a contacts application having a different section for A, B, C through Z. The `numberOfRowsInSection` method returns the number of elements in each section; in this case, as there is only one section, the number of objects in the array.

> The reason why each method is called `tableView` and takes a `tableView` argument is a result of the Objective-C heritage of UIKit. The Objective-C convention combined the method name as the first named argument, so the original method was [delegate tableView:UITableView, numberOfRowsInSection:NSInteger]. As a result, the name of the first argument is reused as the name of the method in Swift.

The `cellForRowAtIndexPath` method is expected to return `UITableViewCell` for an object. In this case, a cell is acquired from the `tableView` using the `dequeueReusableCellWithIdentifier` method (which caches cells as they go off screen to save object instantiation), and then the `textLabel` is populated with the object's `description` (which is a `String` representation of the object; in this case, the date).

This is enough to display elements in the table, but in order to permit editing (or just removal, as in the sample application), there are some additional protocol methods that are required:

```
override func tableView(tableView: UITableView,
  canEditRowAtIndexPath indexPath: NSIndexPath) -> Bool {
    return true
}
override func tableView(tableView: UITableView,
  commitEditingStyle editingStyle: UITableViewCellEditingStyle,
  forRowAtIndexPath indexPath: NSIndexPath) {
    if editingStyle == .Delete {
      objects.removeObjectAtIndex(indexPath.row)
      tableView.deleteRowsAtIndexPaths([indexPath],
        withRowAnimation: .Fade)
    }
}
```

The `canEditRowAtIndexPath` method returns `true` if the row is editable; if all the rows can be edited, then this will return `true` for all the values.

The `commitEditingStyle` method takes a table, a path, and a style, which is an enumeration that indicates which operation occurred. In this case, `UITableViewCellEditingStyle.Delete` is passed in order to delete the item from both the underlying object array and also from the `tableView`. (The enumeration can be abbreviated to `.Delete` because the type of `editingStyle` is known to be `UITableViewCellEditingStyle`.)

# The DetailViewController class

The detail view is shown when an element is selected in the `MasterViewController`. The transition is managed by the storyboard controller; the views are connected with a *segue* (pronounced *seg-way*; the product of the same name based it on the word *segue* which is derived from the Italian word for *follows*).

To pass the selected item between controllers, a property exists in the `DetailViewController` class called `detailItem`. When the value is changed, additional code is run, which is implemented in a `didSet` property notification:

```
class DetailViewController: UIViewController {
  var detailItem: AnyObject? {
    didSet {
      self.configureView()
    }
  }
  ...
}
```

When `DetailViewController` has the `detailItem` set, the `configureView` method will be invoked. The `didSet` body is run after the value has been changed, but before the setter returns to the caller. This is triggered by the `segue` in the `MasterViewController`:

```
class MasterViewController: UIViewController {
  ...
  override func prepareForSegue(
    segue: UIStoryboardSegue, sender: AnyObject?) {
    super.prepareForSegue(segue, sender: sender)
    if segue.identifier == "showDetail" {
      if let indexPath =
        self.tableView.indexPathForSelectedRow() {
        let object = objects[indexPath.row] as! NSDate
        let controller = (segue.destinationViewController
```

```
          as! UINavigationController)
          .topViewController as! DetailViewController
        controller.detailItem = object
        controller.navigationItem.leftBarButtonItem =
          self.splitViewController?.displayModeButtonItem()
        controller.navigationItem.leftItemsSupplementBackButton =
          true
      }
    }
  }
}
```

The `prepareForSegue` method is called when the user selects an item in the table. In this case, it grabs the selected row index from the table and uses this to acquire the selected date object. The navigation controller hierarchy is searched to acquire the `DetailViewController`, and once this has been obtained, the selected value is set with `controller.detailItem = object`, which triggers the update.

The label is ultimately displayed in the `DetailViewController` through the `configureView` method, which stamps the `description` of the object onto the `label` in the center:

```
class DetailViewController {
  ...
  @IBOutlet weak var detailDescriptionLabel: UILabel!
  function configureView() {
    if let detail: AnyObject = self.detailItem {
      if let label = self.detailDescriptionLabel {
        label.text = detail.description
      }
    }
  }
}
```

The `configureView` method is called both when the `detailItem` is changed and when the view is loaded for the first time. If the `detailItem` has not been set, then this has no effect.

The implementation introduces some new concepts, which are worth highlighting:

* The `@IBOutlet` attribute indicates that the property will be exposed in interface builder and can be wired up to the object instance. This will be covered in more detail in *Chapter 4, Storyboard Applications with Swift and iOS*, and in *Chapter 5, Creating Custom Views in Swift*.

- The weak attribute indicates that the property will not store a *strong* reference to the object; in other words, the detail view will not own the object but merely reference it. Generally, all @IBOutlet references should be declared as weak to avoid cyclic dependency references.

- The type is defined as UILabel! which is an *implicitly unwrapped optional*. When accessed, it performs an explicit unwrapping of the optional value; otherwise the @IBOutlet will be wired up as a UILabel? optional type. Implicitly unwrapped optional types are used when the variable is known to never be nil at runtime, which is usually the case for the @IBOutlet references. Generally, all @IBOutlet references should be implicitly unwrapped optionals.

# Summary

This chapter presented two sample iOS applications; one in which the UI was created programmatically, and another in which the UI was loaded from a storyboard. Together with an overview of classes, protocols, and enums, and an explanation of how iOS applications start, this chapter gives a springboard to understand the Xcode templates that are frequently used to start new projects.

The next chapter, *Storyboard Applications with Swift and iOS*, will go into more detail about how storyboards are created and how an application can be built from scratch.

# 4
# Storyboard Applications with Swift and iOS

Storyboards were originally introduced in Xcode 4.2 with iOS 5.0. Storyboards solved the problem of being able to graphically present the flow of screens in an iOS application, and they also provided a way to edit the content of these screens in one place instead of many separate xib files. Storyboards work in the same way with Swift as with Objective-C, and the *Swift and storyboards* section shows how to integrate Swift code with storyboard transitions.

This chapter will present the following topics:

- How to create a storyboard project
- Creating multiple scenes
- Using segues to navigate between scenes
- Writing custom view controllers
- Connecting views to outlets in Swift
- Laying out views with Auto Layout
- Using constraints to build resizable views

# Storyboards, scenes, and segues

By default, Xcode 7 creates a `Main.storyboard` file instead of a `MainWindow.xib` file for newly-created iOS projects. The `UIMainStoryboardFile` key in the `Info.plist` file points to the application's main storyboard name (without the extension). When the application starts up, the `Main.storyboard` file is loaded instead of the `NSMainNib` entry. Prior versions of Xcode allowed developers to opt in or out of storyboards, but with Xcode 7, storyboards are the default and developers cannot easily opt out. It is still possible to use the `xib` files for individual sections of an application or to use them to load custom classes for prototype table cells. In addition, Xcode 7 creates a `LaunchScreen.storyboard` to display as a splash screen (on iOS 8 and higher) while the application is loading, in preference to prerendered screens at fixed resolutions. This allows devices with many different resolutions (including future unannounced ones) to render pixel-perfect splash screens without having to be rendered at different resolutions for each new device size.

A *storyboard* is a collection of *scenes* (separate screens) that are connected with *segues* (pronounced *seg-ways*). Each scene is represented by a *view controller*, which has an associated *view*. Segues transition between different scenes with a customizable user-interface transition, such as a slide or fade, and they can be triggered from a UI control or programmatically.

# Creating a storyboard project

As the default templates with Xcode 7 use storyboards by default, any of the templates will work. In fact, each of the application templates set up a specific type of view controller and template code. The simplest template to work with and customize is the **Single View Application**, which can be selected by navigating to **File** | **New** | **Project...**. Create a project called `Storyboards`, which uses a single-view application, for experimentation with this chapter. (Refer to the *Creating a single view iOS application* section in *Chapter 3*, *Creating an iOS Swift App*, for more details on how to create a new application.)

# Scenes and view controllers

Standard view controllers can be used to build up an application, which includes the following:

- Split views using a `UISplitViewController` class, which can contain any of the following but may not be embedded in any other view controller

- Tabbed views using a `UITabBarController` class, which can contain any of the following but may only be embedded in a split view or used as the root controller

- Navigational controls can be added to existing controllers with a UINavigationController class, which can contain any of the following and may be embedded in any of the preceding or used as a root view controller

- Paginated views using a UIPageViewController class, which provide both sliding and page curling display options

- Tabular views using a UITableViewController class

- Grid views using a UICollectionViewController class

- Audio-visual content using a AVPlayerViewController class

- OpenGL ES content using a GLKViewController class

- Custom controller content using a UIViewController class or a custom subclass

These classes can be mixed, but there is an explicit ordering that must be followed to satisfy the Apple **Human Interface Guidelines** (also known as the **HIG**). These are all optional, but if combined, they need to obey this ordering:

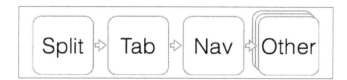

In addition to the standard view controller classes, custom subclasses can be used as well. This is covered in more detail in the *Custom view controllers* section later in this chapter.

# Adding views to the scene

The Main.storyboard file can be opened by clicking on the file in the project navigator. An editor will open, which shows the storyboard as a set of scenes along with the document outline on the left. In a single-page application, only one view controller will exist.

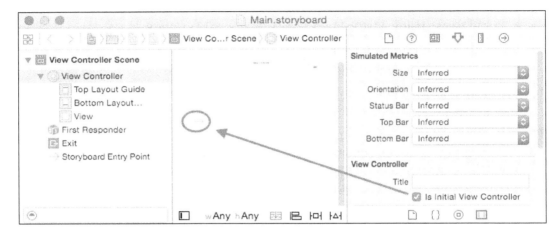

The arrow to the left of the view controller indicates that this scene is the *initial view controller*. This can also be set with the **Is Initial View Controller** checkbox, which can be seen by selecting the **View Controller** from the scene and navigating to the *attributes inspector* (go to **View | Utilities | Show Attributes Inspector**, or press *Command + Option + 4*). The initial view controller can also be changed to a different scene by dragging and dropping the arrow to point to a different scene.

Views are added by dragging and dropping them from the *object library* at the bottom-right of Xcode. The object library can be shown by navigating to **View | Utilities | Show Object Library**, or by pressing *Command + Option + Control + 3*. Click on a view, such as the **Label**, and drag it into the view:

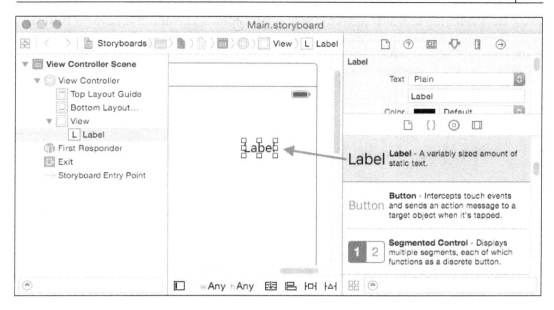

The label's text content can be modified by double-clicking on the label in the view and typing or by selecting the object and editing the text attribute in the attributes inspector:

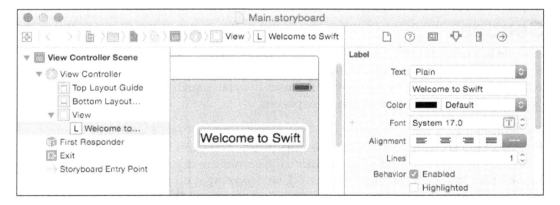

When the element is dragged, blue guide lines may be shown. They suggest locations for the views; the standard is to have a 20pt gap between the views and the edge of the screen and an 8pt gap between adjacent views.

Drag the **Welcome to Swift** label to the top-left of the scene and then drag a **Button** from the object library into the scene. Rename the button's title to **Press Me**. This button should be a standard space (8pt) away from the label and aligned at the baseline (the level at which the text naturally sits).

 At this point, the text in the views is hardcoded in the user interface file and the alignment is manual, which means that the views will not resize if the parent view is modified. These problems will be addressed in the *Connecting views to outlets in Swift* and *Using Auto Layout* sections later in this chapter.

To view the storyboard in the simulator, click on the **Play** button at the top or press *Command + R* to run the application. A window should be shown with **Welcome to Swift** and **Press Me**. At this stage, pressing the button will have no effect, which will be fixed in the next section.

# Segues

A *segue* is a transition to a different scene in a storyboard. Segues can be hooked up to views on the screen or can be triggered via code. The most common transitions are when the user has selected a view in the user interface, such as a button, a table row, or a details icon, and a new scene is displayed.

To demonstrate a segue, a new scene is required. Drag a **View Controller** from the object library and drop it onto the storyboard. The exact location of the view controller doesn't matter, but conventionally, scenes are organized from left to right in the order in which they will be viewed, so dropping it on the right-hand side of the existing view controller is recommended, as shown in the following screenshot:

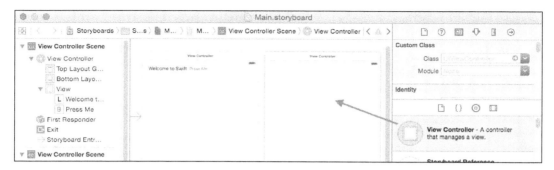

Once the **View Controller** has been added, drop a label onto the top-left and change the text to **Please do not press this button again**. This will present a visual clue that the screen has changed when the segue is followed.

Now, select the **Press Me** button and press the *Control* key while dragging the mouse to the newly created view controller. When the mouse button is released, a pop-up menu will be shown with a number of options that are grouped into **Action Segue** and **Non-Adaptive Action Segue**. The former is preferred; the latter is only there for backward compatibility and might be removed in the future.

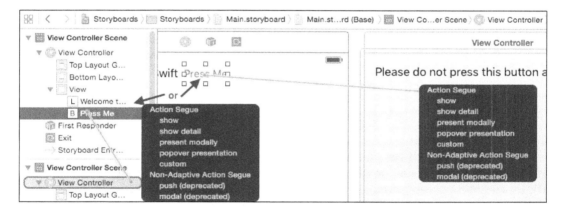

Alternatively, the object can be selected from the *document outline* on the left, and dragged to the object below in the document outline. It is possible to drag from the view in the editor area to an object in the document outline and vice versa. Dragging to the document outline is sometimes faster and more accurate, especially when there are multiple scenes in a storyboard. The document outline can be displayed by navigating to **Editor | Show Document Outline**, if it is not visible, or by clicking on the icon at the bottom-left of the editor.

Choose the **Show** option and a segue will be created between the two views. This is represented as an arrow connecting them and another object in the document outline. The icon inside the circular-segue line shows what kind of transition will occur; a **push** will have an arrow pointing to the left, while **present modally** will be represented as a square box. The **popover** type will show a small popover icon in the segue.

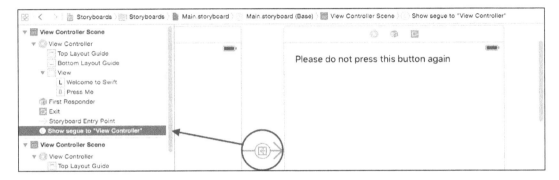

Run the application in the simulator and click the **Press Me** button. A window should slide up and display the second message.

 There will be no way to dismiss or exit the second screen. This is intentional and will be fixed in the next section.

# Adding a navigation controller

When there are multiple screens to be displayed, a parent controller is required to keep track of which screen is currently being shown and what the next step (or previous step) is. This is the purpose of a *navigation controller*; although it has no direct visual representation, it is represented as a scene in a storyboard and can affect the layout of the individual elements in the storyboard.

To embed the initial scene into a navigation controller, select the initial view and navigate to **Editor | Embed In | Navigation Controller**. This will create a new navigation controller view and place it to the left-hand side of the first scene. It will also change the initial view controller to the navigation controller and set up a *relationship segue* with the name **root view controller** between the navigation controller and the first scene that is represented by an icon that is similar to a percent symbol but with the line rotated the other way:

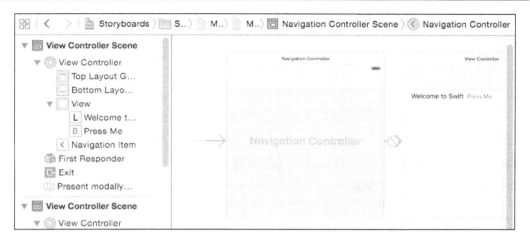

It will be necessary to move the label and button below the newly added navigation bar so that they are still visible. This can either be done before the navigation controller is introduced or by selecting through overlapping objects.

To temporarily hide the navigation bar, delete the relationship segue between the navigation controller and the welcome scene, and the navigation bar will disappear. This will allow the objects to be selected and moved elsewhere temporarily in order to be repositioned. To add it back again, press the *Control* key and drag the mouse cursor from the navigation controller to the welcome scene and choose **root view controller** under **Relationship Segue**; or alternatively, set the **Top Bar** attribute to **None** in the attribute inspector.

Alternatively, to select through overlapping objects, first select the object in the document outline so that the location is shown with the drag boxes. Then, press the *Shift* key and right-click it for a pop-up menu of the objects under the mouse position at any depth. From here, the object can be selected and then moved with the arrow keys to reposition them elsewhere.

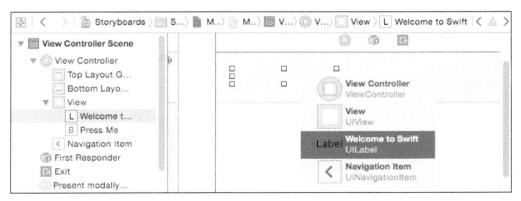

Now when the application is run and the **Press Me** button is tapped, the message will be shown again but with a **< Back** navigation menu item as well, as shown here:

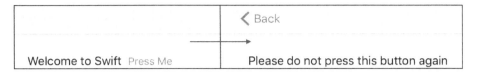

# Naming scenes and views

When working with many scenes, calling all of them **View Controller Scene** is not helpful. To distinguish between them, the controllers can be renamed in the storyboard editor.

To change the name of a scene, select its view controller in the document outline and go to **View | Utilities | Show Attributes Inspector** or press *Command + Option + 3*, and then drill down to the **Document** section where the label hint will read **Document Label**. Typing in another value, such as Press Me, Message, or Initial will rename both the view controller and the scene in the document outline:

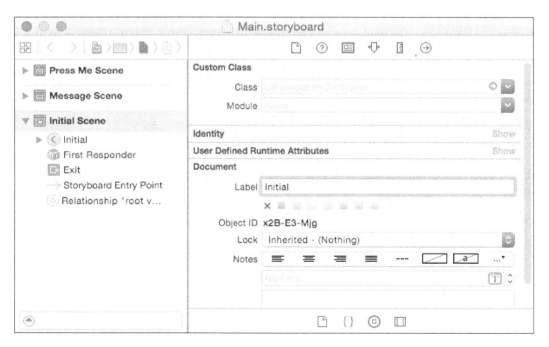

By default, the name of the element in the document outline is taken from the text value of the element or the type if no text value is present. This means that updates to the label or button text will be automatically reflected in the outline. However, it is possible to add document labels to any view in the document outline.

# Swift and storyboards

So far in this chapter, the storyboard content does not involve any Swift or other programming content—it used the drag and drop capabilities of the storyboard editor. Fortunately, it is easy to integrate Storyboard and Swift using a *custom view controller*.

# Custom view controllers

Each standard view controller has a corresponding superclass (listed in the *Scenes and view controllers* section previously in this chapter). This can be replaced with a custom subclass, which then has the ability to influence and change what happens in the user interface. To replace the message in the **Message Scene**, create a new file named `MessageViewCotroller.swift` with the following content:

```
import UIKit
class MessageViewController: UIViewController {
}
```

Having created this class, it can be associated with the view controller by selecting it in the storyboard and then switching to the identity inspector by navigating to **View | Utilities | Show Identity Inspector** or pressing *Command + Option + 3*. In the **Custom Class** section, the **Class** will show `UIViewController` as a hint. Entering `MessageViewController` here will associate the custom controller with the view controller:

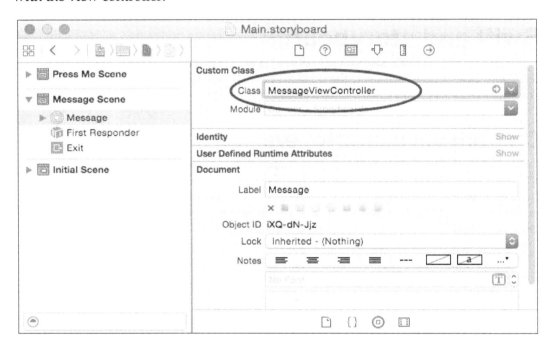

This will have no visible impact to the message scene; running the application will be the same as before. To show a difference, create a `viewDidLoad` method with an `override` keyword and then create a random color for the background as follows:

```
override func viewDidLoad() {
  super.viewDidLoad()
  let red = CGFloat(drand48())
  let green = CGFloat(drand48())
  let blue = CGFloat(drand48())
  view.backgroundColor = UIColor(
    red:red,
    green:green,
    blue:blue,
    alpha:1.0
  )
}
```

Running the application and pressing the **Press Me** button results in a differently colored view being created each time.

> This does not demonstrate good user experience, but is used here to demonstrate the fact that viewDidLoad is called each time the segue occurs. It is typically used to set up view state just before showing the view to the user.

# Connecting views to outlets in Swift

Each view controller has an implicit relationship with its view, and each view has its own backgroundColor property. This example will work regardless of what the view happens to be. What if the view controller needs to interact with the view's content in some way? The view controller could walk the view programmatically, looking for a certain type of view or for a view with a particular identifier, but there is a better way to do this.

Both the interface builder and storyboard have the concept of *outlets*, which are a predefined point in a class that can be exposed and can have connections between the UI and the code. In Objective-C, this was done with an IBOutlet qualifier. In Swift, this is done with a @IBOutlet attribute. In effect, they are variables that can be bound to the UI.

> When defining a class with a @IBOutlet attribute, the @objc attribute is also implicitly added marking this Swift class as using the Objective-C runtime. As all the UIKit classes are already Objective-C types, this doesn't matter; but for types where the Objective-C runtime should not be used, care should be taken when adding attributes, such as @IBOutlet. The @objc attribute can also be used for non-UI classes that need to use the Objective-C runtime.

The following steps are required to create an outlet in a Swift view controller:

1. Define an outlet in the view controller code with @IBOutlet weak var of an optional type of the connected view.
2. Connect the outlet in the view controller to the view by pressing *Control* and dragging the mouse cursor from the view to the outlet.

To do this, open the **assistant editor** by pressing *Command + Option + Enter* or by going to **View | Assistant Editor | Show Assistant Editor**. This will show a side-by-side view of the associated source file. This is useful to display the associated custom view controller for a selected view in the storyboard (or the interface file).

Once the assistant editor is displayed, open the **Message Scene** from the storyboard and press *Control* while dragging the mouse cursor from the message label to the assistant editor and dropping it just after the class declaration:

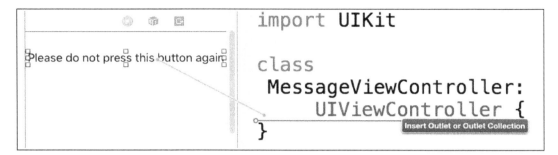

A pop-up dialog will ask what to call the field and present some other information; ensure **Outlet** is selected, name it message, and ensure that it has a **Weak** storage type:

This will result in the following line being added to the MessageViewController class, and it will wire up the label to the property as follows:

```
class MessageViewController: UIViewController {
    @IBOutlet weak var message: UILabel!

    ...
}
```

The @IBOutlet attribute (defined in UIKit) allows interface builder to bind to the property. The **Weak** storage type—which can be changed in the pop-up dialog—indicates that this class will not hold a strong reference to the object so that when the view is dismissed, the controller will not own it.

Generally, all @IBOutlet connections should be marked as weak, because the storyboard or the xib file is the owner of the object, not the controller. Ownership does not pass when assigning properties from interface builder. Changing it to something other than weak may lead to circular references. As Swift uses a reference counting approach to determine when an object is no longer referenced, a circular reference between strong references can cause memory leaks.

The exclamation mark on the end of the type UILabel! indicates that it is an *implicitly unwrapped optional*. This property is stored as an optional type, but the accessor code will automatically unwrap it at the point of use. As the view controller will not have a reference to the message at the point of initialization, it will be nil, so it must be stored as an optional. However, as the value is known to not be nil after the view has been loaded, the implicitly unwrapped optional saves the ?. calls that would otherwise be used each time it is used.

An implicitly unwrapped optional is still an optional value under the covers; it is syntactic sugar to unwrap it at the point of use each time the value is accessed. When the view is loaded, but before the viewDidLoad method is called, the outlet's value will be wired to the instantiated view on screen.

The connections can be seen in the connections inspector, which can be displayed by selecting the message label and pressing *Command + Option + 6* or by navigating to **View | Utilities | Show Connections Inspector**. The inspector can also be used to remove existing connections or add new ones.

Now that the connection has been made between the message view and the custom controller, instead of changing the background color of the view, change the background color of the message instead, as follows:

```
message.backgroundColor = UIColor(...)
```

Run the application and the message will have the background color changed each time the scene is displayed:

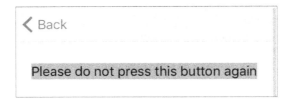

# Calling actions from interface builder

In the same way that outlets are variables for interface builder to assign to (or read from), *actions* are methods/functions that can be triggered from a view in interface builder. The @IBAction attribute is used to annotate a method or function that can be wired up.

 As with @IBOutlet, using @IBAction on a function causes the compiler to implicitly add a @objc attribute to the class in order to force it to use the Objective-C runtime.

To change the message when a button is invoked, a suitable changeMessage is required. Historically, the signature for an action method was one that returned void, marked with IBAction, and took a sender argument, which could be any object. In Swift, this signature translates to the following:

```
@IBAction func changeMessage(sender:AnyObject) { … }
```

However, with Swift, the sender is no longer a required argument. It is, therefore, possible to bind an action with the following signature:

```
@IBAction func changeMessage() { … }
```

If the signature is changed, any existing bindings must be deleted and recreated, as an error will be reported otherwise.

 It is difficult to convert from a func that doesn't take an argument to one that takes an argument. It is easier to have a func that takes an argument that isn't required. If not sure, choose the function signature that takes a sender object and then just ignore it.

The `changeMessage` function can randomly select a message and set the text on the label, as follows:

```
let messages = [
  "Ouch, that hurts",
  "Please don't do that again",
  "Why did you press that?",
]
@IBAction func changeMessage() {
  message.text = messages[
    Int(arc4random_uniform(
      UInt32(messages.count)))]
}
```

When the function is invoked, the message text will change to a value that is defined in the array. To call the function, it needs to be wired up in the storyboard editor. Add a new **Button** from the object library to the message scene, with a `Change Message` label. To connect it to the action, press *Control* and drag the mouse cursor from the **Change Message** button in **Message Scene** and drop it on the **Message** view controller at the top:

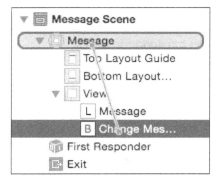

A pop-up menu will then display the outlets and actions that this can be connected to. Select the **changeMessage** from the list:

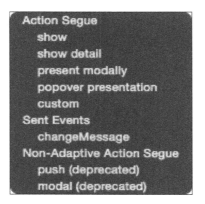

 If **changeMessage** isn't listed, check that the view controller is defined to be MessageViewController and verify that the @IBAction attribute is added to the changeMessage function.

Now when the application is run and the **Change Message** button is pressed, the label will change to one of the hardcoded values.

 The message label will not change in size because the view has no automatic layout associated with it. The *Using Auto Layout* section in this chapter explains how to fix this problem.

# Triggering a segue with code

A segue can be triggered programmatically from code if additional setup is required or if there are data parameters that need to be passed from one view controller to another (such as the currently-selected object).

Segues have named *segue identifiers*, which are used in code to trigger specific segues. To test this out, drag a new **View Controller** from the library (by pressing *Command + Option + Control + 3* or by navigating to **View | Utilities | Show Object Library**) onto the main storyboard and name it About. Drag a **Label** and give it the text: About this App.

Next, create a segue by pressing *Control* and dragging the mouse cursor between the **Message** scene to the new scene. The named identifier can be set as about through the attributes inspector (shown by pressing *Command + Option + 4* or by navigating to **View** | **Utilities** | **Show Attributes Inspector**):

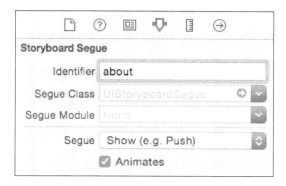

Finally, drag a new **Button** to the **Change Message** scene and call it About. Instead of directly calling the segue, create a new @IBAction called about. When this button is pressed, the following code will be run:

```
@IBAction func about(sender: AnyObject) {
  performSegueWithIdentifier("about", sender: sender)
}
```

When the **About** button is pressed, the **About** screen will be displayed.

# Passing data with segues

Typically, in a master-detail application, data needs to be passed from one scene to the next. This may be the currently selected object, or it may require additional information to be processed. When the segue is called, the view controller's prepareForSegue method is called, with the destination segue and the sending object. This allows any internal state of the view controller to be passed to the new segue.

The `UIStoryboardSegue` contains an identifier, which was set in the previous section. As the `prepareForSegue` method may be called on the `MessageViewController` for any number of segues, it is common for a `switch` statement to be used on the identifier so that the right action can be taken. For a single segue, an `if` statement can be used as follows:

```
override func prepareForSegue(segue: UIStoryboardSegue,
  sender: AnyObject?) {
  if segue.identifier == "about" {
      let dest = segue.destinationViewController as UIViewController
      dest.view.backgroundColor = message.backgroundColor
  }
}
```

Here, the `prepareForSegue` method is called with `segue`, which contains the destination (the scene) and the identifier. The `if` statement ensures that the correct identifier is matched. In this case, the background color of the message label (which is chosen randomly when the view is loaded) is passed to the destination view's background color; however, any property on either the view controller or the view can be set here.

# Using Auto Layout

**Auto Layout** has been part of Xcode for the last few releases, and it was added to support an evolution from the previous springs-and-struts approach that predated Mac OS X. First released on iOS 6.0, it has evolved to the point where size-independent displays can now be created as the default.

## Understanding constraints

In Xcode 5, interface builder enabled Auto Layout by default for the first time. When a label was dragged to the top or bottom of the parent view, a dotted blue line would indicate that the label was correctly spaced, and a *constraint* would be generated.

However, in many cases, the constraints weren't created correctly or had undesired effects. For example, positioning a button in the center at the top may not maintain the location depending on whether the constraint being added was absolute (200px from the right) or relative (in the center of the screen). In both cases, the button may look like it was positioned correctly, only to fail when the device's screen orientation rotates or it is run on a screen of different size.

In Xcode 6, although the guidelines are still displayed as views are moved around, relative constraints are not created. Instead, each view is given an exact hardcoded position that does not change with screen rotation or with a change of display size.

In Xcode 7, Auto Layout is the preferred way of creating applications, and views are implicitly selected for Auto Layout. In addition, separate user interfaces can be created for different *size classes*, which allows applications such as Calculator and Mail to provide different user interfaces that are based on the device's rotation. On larger screen devices that have the ability to dock applications next to each other, the size classes are used to determine how each application looks and behaves.

Constraints must be added manually to the views in order to restore the right behavior, and as manual constraints are added, absolute constraints are removed.

# Adding constraints

In the example application, the **Welcome to Swift** label and the **Press Me** button are next to each other, a small distance from the top. However, when the screen is rotated in the simulator, by pressing *Command* and the left or right arrow keys, the spacing between the labels and the top doesn't change, so the labels look further away.

The desired outcome is that the label remains a standard distance away from the top-left edge and the button remains aligned to the label's baseline.

There are two separate constraints that need to be applied to the label:

* Be a standard vertical distance away from the top of the parent view
* Be a standard horizontal distance away from the left of the parent view

There are also two constraints that need to be applied to the button:

* Be aligned with the label's baseline
* Be a standard vertical distance away from the label

There are different ways of adding a constraint, which are covered in the following sections.

# Adding a constraint with drag and drop

A quick way to add a constraint is to press *Control* and drag the mouse cursor from the view to the top of the container. Depending on the direction of the drag, different options will be displayed. Dragging vertically upwards presents the vertical alignment options:

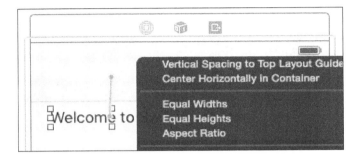

The **Vertical Spacing to Top Layout Guide** option will insert a recommended break between the navigation bar and the label. There is a **Center Horizontally in Container** option, which is also a vertical separation but not appropriate in this case.

The other types that are active—**Equal Widths**, **Equal Heights**, and **Aspect Ratio**—allow multiple views to be sized relative to each other.

Dragging horizontally will show a different set of options at the top, including **Leading Space to Container Margin** and **Center Vertically in Container**:

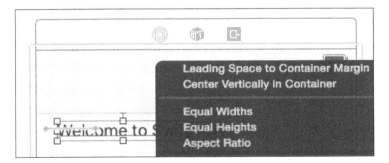

If the mouse is dragged at an angle, both sets of options will be displayed, as follows:

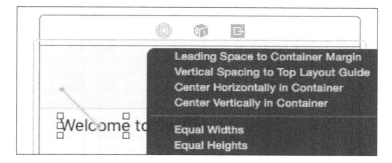

# Adding constraints to the Press Me scene

To set the constraints for the welcome label, press *Control* and drag the mouse cursor from the label to the left, and select **Leading Space to Container Margin**. An orange line will appear, and an orange outline will be displayed:

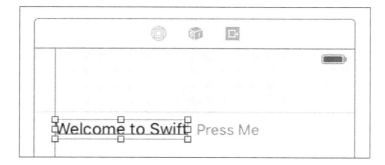

The orange line indicates an *ambiguous constraint*, which means that some constraints have been added to the view but are not enough to uniquely position the label. In this case, the label is positioned from the left of the container, but it could be anywhere with respect to the top or bottom of the screen. The red dotted lines show where the Auto Layout algorithm will place the view with the constraints that are currently specified.

To resolve this problem, press *Control* and drag the mouse pointer from the label to the top and select **Vertical Spacing to Top Layout Guide**. Once this is done, two constraints will be displayed in blue, which represent the constraints about the object:

> If there is an orange box surrounding the label along with a warning that says **Frame for label will be different at run-time**, this can be fixed with the **Update Frames** option that is discussed in the next section.

The constraints can also be seen in the document outline on the left-hand side:

If the application is run now and rotated, the label is correctly repositioned, but the button is not:

# Adding missing constraints

To find out which views have no constraints, click through the views one by one in the document outline and check the size inspector (which can be seen by pressing *Command + Option + 5* or by navigating to **View | Utilities | Show Size Inspector**). For views that have constraints set, there will be content shown under the **Constraints** section:

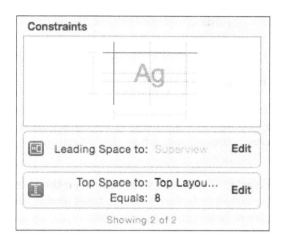

If a view has no constraints associated with it, then this section will be empty. Interface builder has an option to create missing constraints for selected views, which can be accessed by navigating to **Editor | Resolve Auto Layout Issues | Add Missing Constraints** or from the **Resolve Auto Layout Issues** menu at the bottom-right, which looks like a triangle between two vertical lines.

When selected, the options in the top-half apply to selected views only, while the options in the bottom-half work on all the views in the selected view controller:

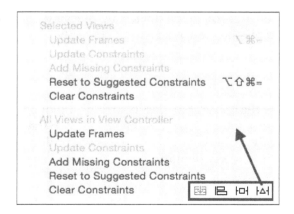

The options include:

- **Update Frames**: This is based on the current constraints; it automatically repositions and resizes the views to correspond to what will happen at runtime
- **Update Constraints**: This is based on the current positions of the objects and attempts to recalculate the existing constraints (but not create new ones)
- **Add Missing Constraints**: This is based on the approximate positioning of the components and adds constraints that creates the same result
- **Reset to Suggested Constraints**: This is equivalent to clearing all the constraints associated with the views and then reading missing constraints
- **Clear Constraints**: This removes all the constraints associated with the views

To add constraints to the **Press Me** button, click on the view and then navigate to **Editor | Resolve Auto Layout Issues | Selected Views | Add Missing Constraints**. There should be two constraints added: a baseline alignment with the label, and a horizontal space to the label.

To see the effect of the **Update Frames** operation, move the label and the button to different places in the view controller. Orange lines and dotted outlines will be shown, indicating that there is an ambiguous constraint. Navigate to **Choose Editor | Resolve Auto Layout Issues | All Views in View Controller | Update Frames**, and the views will automatically move to the right places and resize.

> The views are sized to their *intrinsic size*, which is the size that just fits the content. For example, a label's intrinsic size is the size in which the text can fit into the space in the current font. This can be used to fix the size of the label in the **Message Scene**; by adding constraints, the changing text will result in the intrinsic size being recalculated, and the background color will be correctly sized.

Now, run the application and rotate the device, by pressing *Command* and the left and right arrow keys to see the view resize itself correctly.

# Summary

This chapter introduced the concept of storyboards as a sequence of scenes that are connected with segues, which can either be wired with the GUI or driven programmatically. Finally, Auto Layout can be used to build applications that respond to differences in screen orientation or size, as well as respond to changes in view size or other properties.

The next chapter will present how to create custom views in Swift.

# 5
# Creating Custom Views in Swift

User interfaces can be built by combining standard views and view controllers through Interface Builder, Storyboard Editor, or with custom code. However, it will eventually become necessary to break apart a user interface into smaller, reusable, and easier to test segments. These are known as *custom views*.

This chapter will present the following topics:

- Customizing table views
- Building and laying out custom view subclasses
- Drawing graphical views with drawRect
- Creating layered graphics with animation

## An overview of UIView

All iOS views are rooted in an Objective-C class called UIView, which comes from the UIKit framework/module. The UIView class represents a rectangular space that may be associated with UIWindow or constructed to represent an off-screen view. Views that perform user interactions are generally subclasses of UIControl. Both UIView and UIViewController inherit from the UIResponder class, which in turn inherits from NSObject:

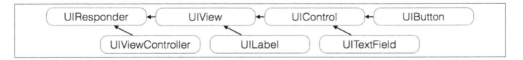

On Mac OS X, views are rooted in NSView and come from the AppKit framework. Otherwise, these two implementations are very similar. A new Xcode project will be used to create custom view classes. Create a new project called CustomViews that is based on the **Tabbed Application** template. To start with a blank sheet, delete the generated view controllers from the Main.storyboard and their associated FirstViewController and SecondViewController classes.

# Creating new views with Interface Builder

The easiest way to create a custom view is to use Interface Builder to drag and drop the contents. This is typically done with a UITableView and a *prototype table cell*.

## Creating a table view controller

Drag in a **Table View Controller** from the object library onto the main storyboard, and drag and drop from the tab bar controller to the newly created table view controller to create a relation segue called view controllers. (Segues are covered in more detail in the *Storyboards, Segues, and Scenes* section in *Chapter 4, Storyboard Applications with Swift and iOS*.)

By default, the table view controller will have *dynamic property content*—that is, it will be able to display a variable number of rows. This is defined in the **Table View** section of **Attributes Inspector**, which can be displayed by selecting **Table View** from the scene navigator and then pressing *Command + Option + 4*:

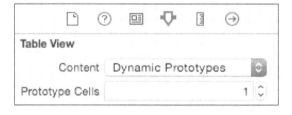

There is an option for tables to have *static content*; a fixed number of rows in the table. This is sometimes useful when creating scrollable content that can be partitioned into slices, even if it doesn't look like a table. Most of the elements in the iOS settings are represented as a fixed-size table view. At the top of the table view are one or more *prototype cells*. These are used to define the look and feel of the table items. By default, a UITableViewCell is used, which has a label and an image, but a prototype cell can be used to add more data to the entries.

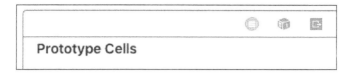

The prototype cell can be used to provide additional information or views. For example, two labels can be dragged into the view; one label can be centered at the top and can be displayed in the headline font, while the second can be left-aligned.

Drag two **UILabels** from the object library into the prototype cell and arrange them using **Auto Layout**, appropriately.

To change a label's font, select the label in the editor and go to **Attributes Inspector**. In the **Label** section, click on the **Font Chooser** icon and select **Headline** or **Subhead**, as appropriate:

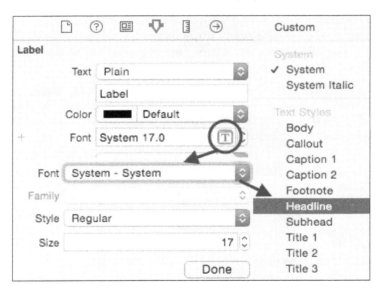

When finished, the prototype cell will look similar to the following screenshot:

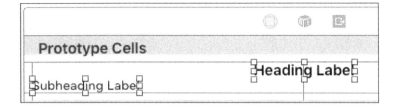

When the application is run, an empty table will be seen. This is because the table doesn't have any items displayed at the moment. The next section shows how to add data to a table so that it binds and displays items to the prototype cell.

# Showing data in the table

A UITableView acquires data from a UITableViewDataSource. The UITableViewController class already implements the UITableViewDataSource protocol, so only a small number of methods are required to provide data for the table.

 As UITableView was originally implemented in Objective-C, the methods that are defined in the protocol take a tableView. As a result, all of the UITableViewDataSource delegate methods in Swift end up being called tableView with different arguments.

Create a new SampleTable class that extends UITableViewController. Implement the class as follows:

```
import UIKit
class SampleTable: UITableViewController {
  var items = [
    ("First", "A first item"),
    ("Second", "A second item"),
  ]
  required init?(coder:NSCoder) {
    super.init(coder:coder)
  }
  override func tableView(tableView: UITableView,
    numberOfRowsInSection section:Int) -> Int {
    return items.count
  }
  override func tableView(tableView: UITableView,
    cellForRowAtIndexPath indexPath: NSIndexPath)
     -> UITableViewCell {
```

```
let cell = tableView.
  dequeueReusableCellWithIdentifier("prototypeCell")!

// configure labels
return cell
}
}
```

Once the data source methods are implemented, the labels need to be configured to display the data from the array. There are three things that need to be done: the prototype cell must be acquired from the `xib` file; the labels need to be extracted; and finally the table view controller needs to be associated with the custom `SampleTable` class.

Firstly, the `cellForRowAtIndex` function needs an identifier for reusable cells. The **Identifier** is set on the prototype cell in the main storyboard. To set this, select the prototype cell and go to the **Attributes Inspector**. Enter `prototypeCell` in the **Identifier** of the **Table View Cell** section:

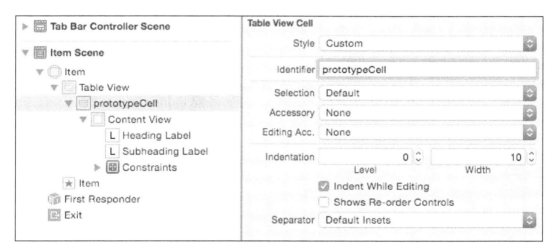

The identifier is used in the `dequeueReusableCellWithIdentifier` method of the `tableView`. When a `xib` is used to load the cell, the return value will either reuse a cell that has gone off screen earlier, or a new cell will be instantiated from the `xib`.

Each label can be given a non-zero integer **Tag** so that the label can be extracted from the prototype cell using the `viewWithTag` method:

```
let titleLabel = cell.viewWithTag(1) as! UILabel
let subtitleLabel = cell.viewWithTag(2) as! UILabel
```

To assign tags to the views, select the **Heading Label**, navigate to **Attributes Inspector**, and change **Tag** to 1. Do the same thing for the **Subheading Label** with **Tag** set to 2:

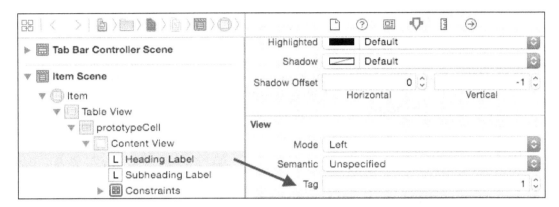

Now, the text values for the row can be set:

```
let (title,subtitle) = items[indexPath.row]
titleLabel.text = title
subtitleLabel.text = subtitle
```

Finally, the `SampleTable` needs to be associated with the table view controller. Click the table, go to **Identity Inspector**, and enter `SampleTable` in the **Custom Class** section:

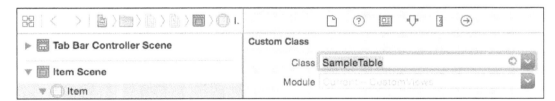

When the application is run, the following view will be displayed:

To hide the status bar, add or change **Status bar is initially hidden** to **YES** and **View controller-based status bar appearance** to **NO** in the `Info.plist` file. Please note that Xcode 7 displays a `CGContextRestoreGState: invalid context 0x0` error message when using these options, which is a known issue that may be fixed in later releases.

# Defining a view in a xib file

It is possible to create a view using **Interface Builder**, save it as a `xib` file, and then instantiate it on demand. This is what happens under the covers with `UITableView`—there is a `registerNib:forCellReuseIdentifier:` method, which takes a `xib` file and an identifier (which corresponds to `prototypeCell` in the previous example).

Create a new interface file named `CounterView.xib` to represent the view, by navigating to **File** | **New** | **File** | **iOS** | **User Interface** | **View**. When opened, it will display as an empty view with no content and in a 600 x 600 square. To change the size to something that is a little more reasonable, go to **Attributes Inspector** and change the size from **Inferred** to **Freeform**. At the same time, change the **Status Bar**, **Top Bar**, and **Bottom Bar** to **None**. Then switch to the **Size Inspector** and modify the view's **Frame Rectangle** to 300 x 50:

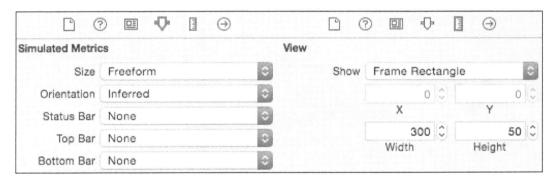

This should resize the view so that it is displayed as 300 by 50 instead of the previous 600 by 600, and the status bar and other bars should not be seen. Now add a **Stepper** from the object library by dragging it to the left-hand side of the view and dragging a **Label** to the right. Adjust the size and add the missing constraints so that the view looks similar to the following screenshot:

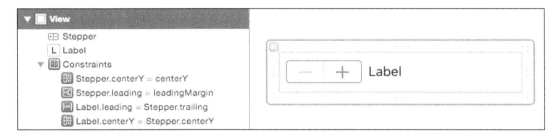

# Wiring a custom view class

Create a new `CounterView` class that extends `UIView`, and define an `@IBOutlet` for the `label` and an `@IBAction change` method that takes a `sender`.

Open the `CounterView.xib` file and select the view. Change **Custom Class** to be `CounterView`. Wire the stepper's `valueChanged` event to the `change` method and connect the `label` outlet:

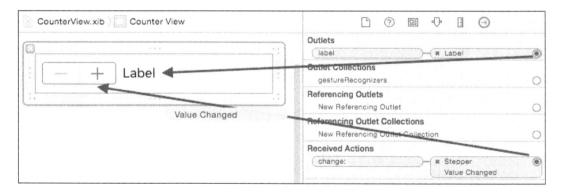

Implement the `change` function such that the label text is changed when the stepper is picked:

```swift
import UIKit
class CounterView: UIView {
  @IBOutlet weak var label:UILabel!
  @IBAction func change(sender:AnyObject) {
```

```
    let count = (sender as! UIStepper).value
    label.text = "Count is \(count)"
  }
}
```

The `CounterView` will be added to the **table header** of the `SampleTable`. Each `UITableViewController` has a reference to its associated `UITableView`, and each `UITableView` has an optional `headerView` (and `footerView`) that is used for the table as a whole.

 The `UITableView` also has `sectionHeader` and `sectionFooter`, which are used to separate different sections of the table. A table can have multiple sections — for example, one section per month — and a separate header and footer can be used per section.

To create a `CounterView`, the `xib` file must be loaded. This is done by instantiating a `UINib` with a `nibName` and a `bundle`. The most appropriate place to do this is in the `viewDidLoad` method of the `SampleTable` class:

```
class SampleTable: UITableViewController {
  override func viewDidLoad() {
    let xib = UINib(nibName:"CounterView", bundle:nil)
    // continued
```

Once the `xib` is loaded, the view must be created. The `instantiateWithOwner` method allows the object(s) in the `xib` to be deserialized.

 It is possible to store multiple objects in a `xib` file (for example, to define a separate view that is suitable for a small display device versus a big display device); but in general, a `xib` file only contains one view.

The owner is passed to the view so that any connections can be wired up to the File's Owner in the interface. This is typically either `self` or `nil` if there are no connections:

```
// continued from before
let objects = xib.instantiateWithOwner(self, options:nil)
// continued
```

This returns an array of `AnyObject` instances, and so casting the first element to a `UIView` is a common step.

 It is possible to use `objects[0]`, but this will cause a failure if the array is empty. Instead, use `objects.first` to get an optional value that contains the first element.

Using the `as?` cast, it is possible to convert the optional value to a more specific type, and from this, perform the assignment to the `tableHeaderView`:

```
// continued from before
let counter = objects.first as? UIView
tableView.tableHeaderView = counter
}
```

When this application is run in the simulator, the following header is seen at the top of the table:

One of the advantages of having a `xib` to represent the user interface is that it can be reused in many places with a single definition. For example, it is possible to use the same `xib` to instantiate another view for the footer of the table, as follows:

```
tableView.tableFooterView =
    xib.instantiateWithOwner(self,options:nil).first as? UIView
```

When the application is run now, counters are created at the top and bottom of the table:

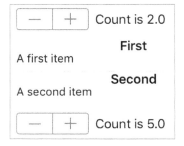

# Dealing with intrinsic size

When a view is added into a view that is being managed with **Auto Layout**, its *intrinsic content size* is used. Unfortunately, views that are defined in **Interface Builder** have no way of setting their intrinsic size programmatically or specifying it in Interface Builder. **Size Inspector** allows this value to be changed, but as Xcode notes, this has no effect at runtime:

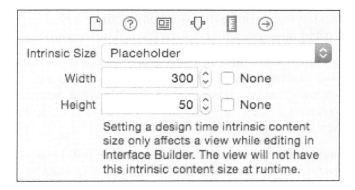

If a custom class is associated with the view, then an appropriate intrinsic size can be defined. Add a method to CounterView that overrides the intrinsicContentSize method and returns a CGSize, allows some xib customization, and returns the maximum of the label's intrinsic size and a value, such as (300,50):

```
override func intrinsicContentSize() -> CGSize {
  let height = max(50,label.intrinsicContentSize().height)
  let width = max(300,label.intrinsicContentSize().width)
  return CGSize(width: width, height: height)
}
```

Now when the view is added into a view that is managed by Auto Layout, it will have an appropriate initial size although it can grow larger.

 The size should take into account the size of the various views that are contained inside, as well as any font sizes or themes, which might change the view. Using the label's intrinsicSize to calculate a maximum is a good idea.

# Creating new views by subclassing UIView

Although the xib files offer a mechanism to customize classes, the majority of UIKit views outside of standard frameworks are implemented in custom code. This makes it easier to reason what the intrinsic size should be as well as to receive code patches and understand diffs from version control systems. The downside of this approach is when using Auto Layout, writing the constraints can be a challenge and the intrinsic sizes are often misreported or return the unknown value: (-1,-1).

A custom view can be implemented as a subclass of UIView. Subclasses of UIView are expected to have two initializers, one that takes a frame:CGRect and one that takes a coder:NSCoder. The frame is generally used in code, and the rect specifies the position on screen (0,0 is the top-left) along with the width and height. The coder is used when deserializing from a xib file.

To allow custom subclasses to either be used in Interface Builder or instantiated from code, it is good practice to ensure that both the initializers create the necessary views. This can be done using a third method called setupView, which is invoked from both.

Create a class called TwoLabels that has two labels in a view:

```
import UIKit
class TwoLabels: UIView {
  var left:UILabel = UILabel()
  var right:UILabel = UILabel()
  required init?(coder:NSCoder) {
    super.init(coder:coder)
    setupView()
  }
  override init(frame:CGRect) {
    super.init(frame:frame)
    setupView()
  }
  // ...
}
```

The setupView call will add the subviews to the view. Code that goes in here should be executed only once. There isn't a standard name, and often, example code will place the setup in one or other of the init methods instead.

It is conventional to have a separate method, such as configureView, to populate the UI with the current set of data. This can be called repeatedly based on the state of the system; for example, a field may be enabled or disabled based on some condition. This code should be repeatable so that it does not modify the view hierarchy:

```
func setupView() {
  addSubview(left)
  addSubview(right)
  configureView()
}
func configureView() {
  left.text = "Left"
  right.text = "Right"
}
```

In an explicitly sized environment (where the text label is being set and placed at a particular location), there is a layoutSubviews method that is called to request the view to be laid out correctly. However, there is a better way to do this, which is to use Auto Layout and constraints.

# Auto Layout and custom views

Auto Layout is covered in the *Using Auto Layout* section of *Chapter 4, Storyboard Applications with Swift and iOS*. When creating a user interface explicitly, views must be sized and managed appropriately. The easiest way to manage this is to use Auto Layout, which requires constraints to be added in order to set up the views.

Constraints can be added or updated in the updateConstraints method. This is called after setNeedsUpdateConstraints is called. Constraints may need to be updated if views become visible or the data is changed. Typically, this can be triggered by placing a call at the end of the setupView method, as follows:

```
func setupView() {
  // addSubview etc
  setNeedsUpdateConstraints()
}
```

The updateConstraints method needs to do several things. To prevent autoresizing masks being translated into constraints, each view needs to call setTranslatesAutoresizingMaskIntoConstraints with an argument of false.

To facilitate the transition between springs and struts (also known as autoresizing masks) and Auto Layouts, views can be configured to translate springs and struts into Auto Layout constraints. This is enabled by default for all views in order to provide backward compatibility for existing views, but it should be disabled when implementing Auto Layouts.

Either the constraints can be incrementally updated or the existing constraints can be removed. A removeConstraints method allows existing constraints to be removed first, as follows:

```
override func updateConstraints() {
  translatesAutoresizingMaskIntoConstraints = false
  left.translatesAutoresizingMaskIntoConstraints = false
  right.translatesAutoresizingMaskIntoConstraints = false
  removeConstraints(constraints)
  // add constraints here
}
```

Constraints can be added programmatically using the NSLayoutConstraint class. The constraints that are added in Interface Builder are also instances of the NSLayoutConstraint class.

Constraints are represented as an equation; properties of two objects are related as an equality (or inequality) of the following form:

```
// object.property = otherObject.property * multiplier + constant
```

To declare that both labels are of equal width, the following can be added to the updateConstraints method:

```
// left.width = right.width * 1 + 0
let equalWidths = NSLayoutConstraint(
  item: left,
  attribute: .Width,
  relatedBy: .Equal,
  toItem: right,
  attribute: .Width,
  multiplier: 1,
  constant: 0)
addConstraint(equalWidths)
```

# Constraints and the visual format language

Although adding individual constraints gives us ultimate flexibility, it can be tedious to set up programmatically. The *visual format language* can be used to add multiple constraints to a view. This is an ASCII-based representation that allows views to be related to each other in position and extrapolated into an array of constraints.

Constraints can be applied horizontally (the default) or vertically. The | character can be used to represent either the start or end of the containing superview, and – is used to represent the space that separates views, which are named in [] and referenced in a dictionary.

To constrain the two labels that are next to each other in the view, H:|-[left]-[right]-| can be used. This can be read as a horizontal (H:) with a gap from the left edge (|-) followed by the left view ([left]), a gap (-), a right view ([right]), and finally, a gap from the right edge (-|). Similarly, vertical constraints can be added with a V: prefix.

The constraintsWithVisualFormat method on the NSLayoutConstraint class can be used to parse visual format constraints. It takes a set of options, metrics, and a dictionary of views that are referenced in the visual format. An array of constraints is returned, which can be passed into the addConstraints method of the view.

To add constraints that ensure the left and right views have equal widths, a space between them, and a vertical space between the top of the view and the labels, the following code can be used:

```
override func updateConstraints() {
    // ...
    let options = NSLayoutFormatOptions()
    let namedViews = ["left":left,"right":right]
    addConstraints(NSLayoutConstraint.
        constraintsWithVisualFormat("H:|-[left]-[right]-|",
            options: options, metrics: nil, views: namedViews))
    addConstraints(NSLayoutConstraint.
        constraintsWithVisualFormat("V:|-[left]-|",
            options: options, metrics: nil, views: namedViews))
    addConstraints(NSLayoutConstraint.
        constraintsWithVisualFormat("V:|-[right]-|",
            options: options, metrics: nil, views: namedViews))
    super.updateConstraints()
}
```

If there are ambiguous constraints, then an error will be printed to the console when the view is displayed. Messages that include the NSAutoresizingMaskLayout constraints indicate that the view has not disabled the automatic translation of the autoresizing mask into the constraints.

# Adding the custom view to the table

The TwoLabels view can be tested by adding it as a footer to the SimpleTable that was created previously. The footer is a special class, UITableViewHeaderFooterView, which needs to be created and added to tableView. The TwoLabels view can then be added to the footer's contentView:

```
let footer = UITableViewHeaderFooterView()
footer.contentView.addSubview(TwoLabels(frame:CGRect.zero))
tableView.tableFooterView = footer
```

Now when the application is run in the simulator, the custom view will be seen:

# Custom graphics with drawRect

Subclasses of UIView can implement their own custom graphics by providing a **drawRect** method that implements the custom drawing routines. The drawRect method takes a CGRect argument, which indicates the area to draw in. However, the actual drawing commands are performed on a Core Graphics context, which is represented by the CGContext class and can be obtained by a call to UIGraphicsGetCurrentContext.

The Core Graphics context represents a drawable area in iOS, and it is used to print as well as draw graphics. Each view has the responsibility to draw itself; the rectangle will either be the full area (for example, the first time that a view is drawn) or it may be a subset of the area (for example, when a dialog has been displayed and then subsequently removed).

*Core Graphics* is a C-based interface (rather than Objective-C-based), so the API is exposed as a set of functions beginning with the `UIGraphics` prefix. As with other drawing APIs, the program can set the current drawing color, draw lines, set a fill color, fill rectangles, and so on.

To test this, create a class called `SquaresView` that is a subclass of `UIView` in a new Swift file.

All views have the standard `init` methods; delegate them to the superclass's implementation. Finally, create a `drawRect` method that takes a `CGRect`. This will be where the custom drawing occurs. The skeleton will look like the following:

```
import UIKit
class SquaresView: UIView {
  required init?(coder: NSCoder) {
    super.init(coder:coder)
    setupView()
  }
  override init(frame: CGRect) {
    super.init(frame:frame)
    setupView()
  }
  func setupView() {
  }
  override func drawRect(rect: CGRect) {
    // drawing code goes here
  }
}
```

Open the `Main.storyboard`, drag in another `UIViewController` and set the custom class of the view to `SquaresView` in **Identity Inspector**. Drag in a relationship segue between the tabbed view controller and the new view controller, and set the tab bar item to `Squares` which will allow testing to move to a different view. If the application is run, a blank view will be seen in the **Squares** tab.

# Drawing graphics in drawRect

To draw graphics in the view, it is necessary to acquire a `CGContext` and then set a drawing (stroke) color. A `UIColor` can be acquired and then converted into a `CGColor` to be able to set it on the graphics context.

Finally, a rectangle can be drawn with CGContextStrokeRect:

```
override func drawRect(rect: CGRect) {
  let context = UIGraphicsGetCurrentContext()
  let red = UIColor.redColor().CGColor
  CGContextSetStrokeColorWithColor(context, red)
  CGContextStrokeRect(context,
    CGRect(x:50, y:50, width:100, height:100))
}
```

When this is run in the simulator, a red rectangle will be displayed on the **Squares** tab.

To draw a green square with a black outline in the middle requires a filled green square to be drawn first, followed by a black square afterwards. (Drawing them in the opposite order will result in the solid green square obliterating the black square.)

There are two different colors in a Core Graphics context: the *stroke color*, which is used to draw lines and paths, and the *fill color*, which is used when creating a filled path. Although the CGContextSetFillColorWithColor function exists, in Swift, there is an easier way of setting this directly with UIColor using the setFill or setStroke methods. The following code will create the green square with a black border:

```
UIColor.greenColor().setFill()
UIColor.blackColor().setStroke()
CGContextFillRect(context,
  CGRect(x:75, y:75, width:50, height:50))
CGContextStrokeRect(context,
  CGRect(x:75, y:75, width:50, height:50))
```

Now when the application is run, the following will be seen:

# Responding to orientation changes

When the screen rotates, the view is stretched and squashed, resulting in the square turning into a rectangle. The drawRect call is not called when the view changes orientation; the existing display is squashed and stretched automatically.

To prevent this, the *content mode* of the view can be changed. There is a UIViewContentMode enumeration that can be specified to cause different behaviors. Using Redraw will result in the drawRect being called when the orientation changes or when the bounds changes size.

 The other enum values are documented in the UIViewContentMode type, and they include scaling options as well as being centered or attached to one of the edges or corners.

The squares can be centered on the screen; instead of starting at the position 50,50, the view's center property can be accessed to find out what the position is. Modify the code as follows:

```
func setupView() {
  contentMode = .Redraw
}
override func drawRect(rect: CGRect) {
  let context = UIGraphicsGetCurrentContext()
  let red = UIColor.redColor().CGColor
  CGContextSetStrokeColorWithColor(context,red)
  CGContextStrokeRect(context,
    CGRect(x:center.x-50, y:center.y-50, width:100, height:100))
  UIColor.greenColor().setFill()
  UIColor.blackColor().setStroke()
  CGContextFillRect(context,
    CGRect(x:center.x-25, y:center.y-25, width:50, height:50))
  CGContextStrokeRect(context,
    CGRect(x:center.x-25, y:center.y-25, width:50, height:50))
}
```

Now when the application is run, the squares will be centered on the screen. If the screen rotates, drawRect will be invoked again and the display will be redrawn.

# Custom graphics with layers

Drawing graphics by overriding `drawRect` is not very performant because all the drawing routines are executed on the CPU. Offloading the graphics drawing to the GPU is both more performant and more power efficient.

iOS has a concept of layers, which are Core Graphics optimized drawing contents. Operations composed on a *layer*, including adding a *path*, can be translated into code that can execute on the GPU and be rendered efficiently. In addition, Core Animation can be used to animate changes on layers efficiently. *Core Animation* is provided in the **QuartzCore** framework/module; the two terms are interchangeable. It is more generally known as Core Animation.

The download progress icon on iOS can be recreated as a `ProgressView` containing layers for the circular outline, a layer for the square stop button in the middle, and a layer for the progress arc. The final view will composite these three layers together to provide the finished view.

Every `UIView` has an implicit associated layer, which can have sublayers added to it. As with views, newly-added layers overlay existing layers. There are several **core animation layer** classes that can be used, which are subclasses of `CALayer`, and they are as follows:

- The `CAEAGLLayer` class provides a way to embed OpenGL content into a view
- The `CAEmitterLayer` class provides a mechanism to generate emitter effects, such as smoke and fire
- The `CAGradientLayer` class provides a way to create a background with a gradient color
- The `CAReplicatorLayer` class provides a means to replicate the existing layers with different transformations, which allows effects, such as reflections and coverflow, to be displayed
- The `CAScrollLayer` class provides a way to perform scrolling
- The `CAShapeLayer` class provides a means to draw and animate a single path
- The `CATextLayer` class allows text to be displayed
- The `CATiledLayer` class provides a means to generate tiled content at different zoom levels, such as a map
- The `CATransformLayer` class provides a means to transform layers into 3D views, such as a coverflow style image animation

# Creating a ProgressView from layers

Create another view class called `ProgressView` which extends `UIView`. Set it up with the default `init` methods, a `setupView`, and a `configureView` method:

```
import UIKit
class ProgressView: UIView {
  required init?(coder: NSCoder) {
    super.init(coder:coder)
    setupView()
  }
  override init(frame: CGRect) {
    super.init(frame:frame)
    setupView()
  }
  func setupView() {
    configureView()
  }
  func configureView() {
  }
}
```

Create a new `Layers Scene` in the `Main.storyboard` by dragging a `UIViewController` from the object library onto the storyboard. Connect it to the tab-bar controller by dragging a relationship segue to the newly created layers view controller. Add the `ProgressView` by dragging a **View** from the object library and giving it a **Custom Class** of `ProgressView`. Size it with an approximate location of the middle of the screen.

Now add an instance variable to the `ProgressView` class called `circle` and create a new instance of `CAShapeLayer`. In `setupView`, set `strokeColor` as `black` and `fillColor` as `nil`. Finally, add the `circle` layer to the view's layer so that it is displayed:

```
let circle = CAShapeLayer()
func setupView() {
  circle.strokeColor = UIColor.blackColor().CGColor
  circle.fillColor = nil
  self.layer.addSublayer(circle)
  configureView()
}
```

`CAShapeLayer` has a `path` property, which is used to perform all the drawing. The easiest way to use this is to create a `UIBezierPath` and then use the `CGPath` accessor to convert it to a `CGPath`.

A *bezier curve* is a way of representing a smooth curve between two points and one or more additional control points. These can be scaled accurately and are easy to compute in a graphics card. A `UIBezierPath` provides a way to represent one or several bezier paths together, resulting in smooth and efficient curve generation.

Unlike the `UIGraphics*` methods, there are no separate `draw*` and `fill*` operations; instead, either the `fillColor` or `strokeColor` is set and then the path is filled or stroked (drawn). The `UIBezierPath` can be constructed by adding segments, but there are several initializers that can be used to draw specific shapes. For example, circles can be drawn with the `ovalInRect` initializer:

```
func configureView() {
  let rect = self.bounds
  circle.path = UIBezierPath(ovalInRect: rect).CGPath
}
```

Now when the application is run, a small black circle will be seen on the **Layers** tab:

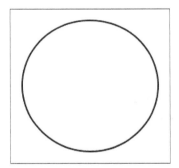

# Adding the stop square

The stop square can be added by creating another layer. This will allow the stop button to be turned on or off as necessary. (For example, during a download, the stop button can be displayed, and when the download is completed, it can be animated away.)

Add a new constant called `square` of type `CAShapeLayer`. It will help to create a constant, `black`, as it will be used again elsewhere in this class:

```
class ProgressView: UIView {
  let square = CAShapeLayer()
  let circle = CAShapeLayer()
  let black = UIColor.blackColor().CGColor
}
```

The `setupView` method can now be updated to deal with additional layers. As it is common to set them up in the same way, using a loop is a quick way to set up multiple layers, as follows:

```
func setupView() {
  for layer in [square, circle] {
    layer.strokeColor = black
    layer.fillColor = nil
    self.layer.addSublayer(layer)
  }
  configureView()
}
```

The path for the `square` can be created using the `rect` initializer of `UIBezierPath`. To create a rectangle that will be centered inside the circle, use the `insetBy` method with an appropriate value:

```
func configureView() {
  let rect = self.bounds
  let sq = rect.insetBy(dx: rect.width/3, dy: rect.height/3)
  square.fillColor = black
  square.path = UIBezierPath(rect: sq).CGPath
  circle.path = UIBezierPath(ovalInRect: rect).CGPath
}
```

Now when the application is run, the following will be seen:

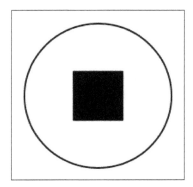

# Adding a progress bar

The progress bar can be drawn as an arc representing the amount of data that has downloaded so far. On other iOS applications, the progress bar starts at the 12 o'clock position and then moves clockwise.

There are two ways to achieve this: using an arc that is drawn up to some particular amount, or by setting a single path that represents the entire circle and then using strokeStart and strokeEnd to define which segment of the path should be drawn. The advantage of using strokeStart and strokeEnd is that they are *animatable properties*, which allow some animated effects.

The arc needs to be drawn from the top, moved clockwise to the right, and then back up again. The strokeStart and strokeEnd are CGFloat values between 0 and 1, so they can be used to represent the progress of the download.

**Easy as Pi**

Although circles are often split into 360 degrees (mainly because 360 has a lot of factors and is easily divisible into different numbers), computers tend to work in *radians*. There are 2pi radians in a circle; so half a circle is pi, and a quarter of a circle is pi/2.

There is a UIBezierPath convenience initializer that can draw an arc; the center and radius are specified along with a startAngle and endAngle point. The start and end points are both specified in radians, with 0 being the 3 o' clock position and going clockwise or anticlockwise as specified:

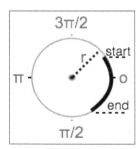

To draw progress starting from the top of the circle, the start point must be specified as -pi/2. Drawing clockwise from here around the complete circle takes it to -pi/2 + 2pi, which is 3 * pi/2.

Computers use pi a lot, defined in `usr/include/math.h`, which is included transitively from `UIKit` through the `Darwin` module. The constants: `M_PI`, `M_PI_2` (pi/2), and `M_PI_4` (pi/4), and the inverses: `M_1_PI` (1/pi), and `M_2_PI` (2/pi), are available.

The middle of the diagram can be calculated by accessing `self.center`, and the radius of the circle will be half the minimum `width` or `height`. To add the path, create a new `CAShapeLayer` called `progress`, add it into the layers array, and optionally give it a different `width` and `color` to distinguish it from the background:

```
class ProgressView: UIView {
  let progress = CAShapeLayer()
  var progressAmount: CGFloat = 0.5
  ...
  func setupView() {
    for layer in [progress, square, circle] {
      ...
    }
    progress.lineWidth = 10
    progress.strokeColor = UIColor.redColor().CGColor
    configureView()
  }
  func configureView() {
    ...
    let radius = min(rect.width, rect.height) / 2
    let center = CGPoint(x:rect.midX, y:rect.midY)
    progress.path = UIBezierPath(
      arcCenter: center,
      radius: radius,
      startAngle: CGFloat(-M_PI_2),
      endAngle: CGFloat(3*M_PI_2),
      clockwise: true
    ).CGPath
    progress.strokeStart = 0
    progress.strokeEnd = progressAmount
  }
}
```

When this is run, the progress bar will be seen behind the circle:

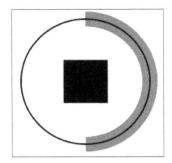

# Clipping the view

The problem with the progress line is that it extends beyond the circular boundary of the progress view. A simple approach may be to try and calculate a half-width distance from the radius and redraw the circle, but this is fragile as changes to the line width may result in the diagram not looking right in the future.

A better approach is to *mask* the graphics area so that the drawing does not go outside a particular shape. By specifying a mask, any drawing that occurs within the mask is displayed; graphics that are drawn outside the mask are not displayed.

A mask can be defined as a rectangular area or the result of a filled layer. Creating a circular mask requires creating a new mask layer and then setting a circular path as we did before.

 A mask can only be used by a single layer. If the same mask is needed for more than one layer, either the mask layer needs to be copied or the mask can be set on a common parent layer.

Create a new CAShapeLayer that can be used for the mask, and create a path that is based on the UIBezierPath with an ovalInRect. The mask can then be assigned to the mask layer of the progress layer:

```
class ProgressView: UIView {
  let mask = CAShapeLayer()
  func configureView() {
    ...
    mask.path = UIBezierPath(ovalInRect:rect).CGPath
    progress.mask = mask
  }
}
```

Now when the display is shown, the progress bar does not bleed over the edge:

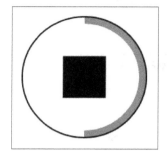

# Testing views in Xcode

To test the view in Interface Builder directly, the class can be marked as `@IBDesignable`. This gives permission for Xcode to instantiate and run the view as well as update it for any changes that are made. If the class is marked as `@IBDesignable`, then Xcode will attempt to load the view and display it in storyboard and `xib` files.

However, when the class loads, the UI will not be displayed properly, because the frame size needs to be initialized correctly. Override the `layoutSubviews` method to call `configureView`, which ensures that the view is properly redrawn when the view changes size or is displayed for the first time:

```
@IBDesignable class ProgressView: UIView {
  ...
  override func layoutSubviews() {
    setupView()
  }
}
```

Now when the `ProgressView` is added or displayed in Interface Builder, it will be rendered in place. Build the project, then open the `Main.storyboard`, and click on the **Progress View**; after a brief delay, it will be drawn.

Xcode can also be used to edit different attributes of an object in Interface Builder. This allows the view to be tested without running the application.

To allow Interface Builder to edit properties, they can be marked as `@IBInspectable`:

```
@IBDesignable class ProgressView: UIView {
  @IBInspectable var progressAmount: CGFloat = 0.5
  ...
}
```

After building the project, open the storyboard, select **Progress View** and go to **Attributes Inspector**. Just above the **View** section will be a **Progress View** section with the **Progress Amount** field that is based on the @IBInspectable field of the same name:

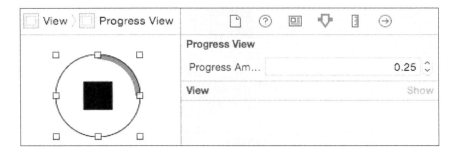

# Responding to change

If UISlider is added to **Layers View**, changes can be triggered by adding @IBAction to allow the valueChanged event to propagate the value to the caller.

Create an @IBAction function, setProgress, which takes a sender and then, depending on the type of that sender, extracts a value:

```
@IBAction func setProgress(sender:AnyObject) {
  switch sender {
    case let slider as UISlider: progressAmount =
      CGFloat(slider.value)
    case let stepper as UIStepper: progressAmount =
      CGFloat(stepper.value)
    default: break
  }
}
```

> Using a switch statement that is based on the type allows additional views to be added in the future.

The valueChanged event on UISlider can now be connected to setProgess on ProgressView.

Assigning the `progressAmount` value alone has no visible effect, so a property observer can be used to trigger display changes whenever the field is modified. A *property observer* is a block of code that gets called before (`willSet`) or after (`didSet`) a property is changed:

```
@IBInspectable var progressAmount: CGFloat = 0.5 {
  didSet {
    setNeedsLayout()
  }
}
```

Now when the application is run and the slider value is moved, the download amount will be updated in the view:

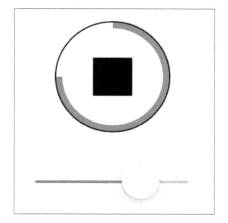

 If the image doesn't update when the slider changes value, check that `didSet` on `progressAmount` triggers a `setNeedsLayout` call, and that the `layoutSubviews` function correctly calls `configureView`.

Observe that the changes to `progressAmount` are animated automatically, so if the slider is quickly moved from one end to the other the download arc will smoothly animate.

 The property observer uses `setNeedsLayout` to trigger a call to `layoutSubviews` in order to achieve the change in display. As changes only need to be picked up when a size change occurs or when a property is changed, this is more efficient than implementing other methods, such as `drawRect`, which will be called every time the display needs to be updated.

# Summary

In this chapter, we looked at several different ways to create views in iOS. The first approach was to use Interface Builder to build the view graphically and analyze some of the problems that this can cause. We then looked at subclassing `UIView` and adding other views to build up a custom view. Finally, we presented two different ways of drawing custom graphics; first with `drawRect`, and subsequently, with layers. The next chapter will show you how to use networking APIs in iOS to download networked data.

# 6
# Parsing Networked Data

Many iOS applications need to communicate with other servers or devices. This chapter presents both HTTP and non-HTTP networking in Swift, and how data can be parsed from either JSON or XML. It first demonstrates how to load data efficiently from URLs, followed by how to stream larger data responses. It then concludes with how to perform both synchronous and asynchronous network requests over protocols other than HTTP.

This chapter will present the following topics:

- Loading data from URLs
- Updating the user interface from a background thread
- Parsing JSON and XML data
- Stream-based connections
- Asynchronous data communication

## Loading data from URLs

The most common way to load data from a remote network source is to use an HTTP (or HTTPS) URL of the form `https://raw.githubusercontent.com/alblue/com.packtpub.swift.essentials/master/CustomViews/CustomViews/SampleTable.json`.

URLs can be manipulated with the `NSURL` class, which comes from the `Foundation` module (which is transitively imported from the `UIKit` module). The main `NSURL` initializer takes a `String` initializer with a full URL, although other initializers exist to create relative URLs or for references to file paths.

The NSURLSession class is typically used to perform operations with URLs, and individual sessions can be created through the initializer or the standard **shared session** can be used. The NSURLConnection class was used in older versions of iOS and Mac OS X. References to this class can still be seen in some tutorials, or may be required if Mac OS X 10.8 or iOS 6 needs to be supported; otherwise, the NSURLSession class should be preferred.

The NSURLSession class provides a means to create tasks. These include:

- **Data task:** This can be used to process network data programmatically
- **Upload task:** This can be used to upload data to a remote server
- **Download task:** This can be used to download to local storage or to resume a previous or partial download

Tasks are created from the NSURLSession class methods, and can take a URL argument and an optional *completion handler*. A completion handler is a lot like a delegate, except that it can be customized per task, and it is usually represented as a function.

Tasks can be *suspended* or *resumed* to stop and start the process. Tasks are created in a suspended state by default, and so they have to be initially resumed to start processing.

When a data task completes, the completion handler is called back with three arguments: an NSData object that represents the returned data, an NSURLResponse object that represents the response from the remote URL server, and an optional NSError object if anything failed during the request.

With this in place, the SampleTable that was created in the previous chapter can load data from a network URL by obtaining a session, initiating a data task, and then resuming it. The completion handler will get called when the data is available, which can be used to add the content to the table.

Modify the viewDidLoad method of the SampleTable class to load the SampleTable.json file by adding the following to the end of the method:

```
let url = NSURL(string: "https://raw.githubusercontent.com/
  alblue/com.packtpub.swift.essentials/master/
  CustomViews/CustomViews/SampleTable.json")!
let session = NSURLSession.sharedSession()
let encoding = NSUTF8StringEncoding
let task = session.dataTaskWithURL(url,completionHandler:
 {data,response,error -> Void in
  let contents = String(data:data!,encoding:encoding)!
  self.items += [(url.absoluteString,contents)]
```

```
// table data won't reload - needs to be on ui thread
    self.tableView.reloadData()
})
task.resume()
```

This creates an NSURL and an NSURLSession, and then creates a data, task and immediately resumes it. After the content is downloaded, the completion handler is called, which passes the data as an NSData object. The String initializer is used to decode UTF8 text from the NSData object, and is explicitly cast to a String so that it can be added to the items array.

 The NSURLSession class also provides other factory methods, including one that takes a configuration argument that includes options, such as whether responses should be cached, whether network connections should go over the cellular network, and whether any cookies or other headers should be sent with the task.

Finally, the item is added to the items and the tableView is reloaded to show the new data. Please note that this does not work immediately if it is not run on the main UI thread; the table has to be rotated or moved in order to redraw the display. Running on the UI thread is covered in the *Networking and user interface* section later in this chapter.

# Dealing with errors

Errors are a fact of life, especially on mobile devices with intermittent connectivity. The completion handler is called with a third argument, which represents any error raised during the operation. If this is nil, then the operation was a success; if not, then the localizedDescription property of the error can be used to notify the user.

For testing purposes, if an error is detected add the localizedDescription to the items in the list. Modify the viewDidLoad method as follows:

```
let task = session.dataTaskWithURL(url, completionHandler:
  {data,response,error -> Void in
    if error == nil {
      let contents = String(data:data!,encoding:encoding)!
      self.items += [(url.absoluteString,contents)]
    } else {
      self.items += [("Error",error!.localizedDescription)]
    }
    // table data won't reload - needs to be on UI thread
    self.tableView.reloadData()
})
```

An error can be simulated using a nonexistent hostname or an unknown protocol in the URL.

# Dealing with missing content

Errors are reported if the remote server cannot be contacted, such as when the hostname is incorrect or the server is down. If the server is operational, then an error will not be reported; but it is still possible that the file that is requested will not be found, or that the server will experience an error while serving the request. These are reported with HTTP status codes.

 If an HTTP URL is not found, the server sends back a 404 status code. This can be used by the client to determine whether a different file should be accessed or whether another server should be queried. For example, browsers will often ask the server for a favicon.ico file and use this to display a small logo; if this file is missing, then a generic page icon is displayed instead. In general, 4xx responses are client errors, while 5xx responses are server errors.

The NSURLResponse object doesn't have the concept of an HTTP status code, because it can be used for any protocol, including ftp. However, if the request used HTTP, then the response is likely to be HTTP and so it can be cast to an NSURLHttpResponse, which has a statusCode property. This can be used to provide more specific feedback when the file is not found. Modify the code as follows:

```
if error == nil {
  let httpResponse = response as! NSHTTPURLResponse
  let statusCode = httpResponse.statusCode
  if (statusCode >= 400 && statusCode < 500) {
    self.items += [("Client error \(statusCode)",
     url.absoluteString)]
  } else if (statusCode >= 500) {
    self.items += [("Server error \(statusCode)",
     url.absoluteString)]
  } else {
    let contents = String(data:data!,encoding:encoding)!
    self.items += [(url.absoluteString,contents)]
  }
} else {...}
```

Now, if the server responds but indicates that either the client made a bad request or the server suffered a problem, the user interface will be updated appropriately.

# Nested if and switch statements

Sometimes, the error handling logic can get convoluted with handling different cases, particularly if there are different values that need to be tested. In the previous section, both the NSError and HTTP statusCode needed to be checked.

An alternative approach is to use a switch statement with where clauses. These can be used to test multiple different conditions and also show which part of the condition is being tested. Although a switch statement requires a single expression, it is possible to use a *tuple* to group multiple values into a single expression.

Another advantage of using a tuple is that it permits the cases to be matched on types. In the networking case, some URLs are based on http or https, which means that the response will be an NSHTTPURLResponse type. However, if the URL is a different type (such as a file or ftp protocol), then it will be of a different subtype of NSURLResponse. Unconditionally casting to NSHTTPURLResponse with as will fail in these cases and cause a crash.

The tests can be rewritten as a switch block as follows:

```
switch (data,response,error) {
  case (_,_,let e) where e != nil:
    self.items += [("Error",e.localizedDescription)]
  case (_,let r as NSHTTPURLResponse,_)
   where r.statusCode >= 400 && r.statusCode < 500:
    self.items += [("Client error \(r.statusCode)",
     url.absoluteString)]
  // see note below
  case (_,let r as NSHTTPURLResponse,_)
   where r.statusCode >= 500:
    self.items += [("Server error \(r.statusCode)",
     url.absoluteString)]
  default:
    let contents = String(data:data!,encoding:encoding)!
    self.items += [(url.absoluteString,contents)]
}
```

In this example, the default block is used to execute the success condition, and the prior case statements are used to match the error conditions.

The case `(_,_,let e) where e != nil` case is an example of a *conditional pattern match*. The underscore, which is called a *wildcard pattern* in Swift (also known as a **hole** in other languages), is something that will match any value. The third parameter, `let e`, is a *value binding pattern*, and has the effect of `let e = error` in this case. Finally, the `where` clause adds the test to ensure this case only occurs when e is not `nil`.

It would be possible to use the identifier `error` instead of `let e` in the `case` statement, using `case (_,_,_) where error != nil` would have had the same effect. However, it is bad practice to capture values outside of the `switch` statement for case matching purposes because if the `error` variable is renamed, then the `case` statement may become invalid. Generally, use `let` patterns inside `case` statements to ensure that the correct expression value is being matched.

The second and third cases perform both a `let` assignment and a type test/conversion. When `case (_,let r as NSHTTPURLResponse,_)` is matched, not only is the value of that part in the tuple assigned the constant `r`, but it is also cast to an `NSHTTPURLRepsonse`. If the value is not of type `NSHTTPURLResponse`, then the case statement is automatically skipped. This is equivalent to an `if` test with an `is` expression followed by a cast with `as`.

Although the patterns are the same in both, the `where` clauses are different. The first `where` clause looks for the case where `r.statusCode` is 400 or greater and less than 500, while the second is matched where `r.statusCode` is 500 or greater.

Whether nested `if` statements or the `switch` statement is used, the code that performs the test is likely to be very similar. It typically comes down to developer preference, but more developers are likely to be familiar with nested `if` statements. In Swift, the `switch` statement is more powerful than in other languages, and so, this kind of pattern is likely to become more popular.

An alternative with Swift 2 is to use the `guard` statement to ensure that if certain error conditions occur, then appropriate action can be taken instead. The `guard` statement is like an `if` statement where there is no `true` block and the `false` block must always leave the function. For example, the code could be rewritten as:

```
guard error == nil else {
  self.items += [("Error",error!.localizedDescription)]
  return
}
```

```
let statusCode = (response as! NSHTTPURLResponse).statusCode

guard statusCode < 500 else {
  self.items += [("Server error \(statusCode)",
    url.absoluteString)]
  return
}

guard statusCode < 400 else {
  self.items += [("Client error \(statusCode)",
   url.absoluteString)]
  return
}

let contents = String(data:data!,encoding:encoding)!
self.items += [(url.absoluteString,contents)]
```

Please note that the guard block must exit the calling function; so, if additional operations are required, either the body of the implementation must be moved to a different function or the switch or if blocks used instead. The examples later in this chapter assume the use of the if blocks for simplicity.

# Networking and user interfaces

One outstanding problem with the current callback approach is that the callback cannot be guaranteed to be called from the *main thread*. As a result, user interface operations may not work correctly or throw errors. The right solution is to set up another call using the main thread.

Accessing the main thread in Swift is done in the same way as it is in Objective-C: using **Grand Central Dispatch** (**GCD**). The *main queue* can be accessed with dispatch_get_main_queue, which is used by the thread that all UI updates should use. Background tasks are submitted with dispatch_async to a queue. To invoke the reloadData call on the main thread, wrap it as follows:

```
dispatch_async(dispatch_get_main_queue(), {
  self.tableView.reloadData()
})
```

This style of call will be valid for both Objective-C and Swift (although Objective-C uses the ^ (caret) as a block prefix). However, Swift has a special syntax for functions that take blocks; the block can be promoted out of the function's argument and left as a trailing argument. This is known as a *trailing closure*:

```
dispatch_async(dispatch_get_main_queue()) {
   self.tableView.reloadData()
}
```

Although this is a minor difference, it makes it look like `dispatch_async` is more like a keyword, such as `if` or `switch`, which takes a block of code. This can be used for any function whose final argument is a function; there is no special syntax needed in the function definition. Additionally, the same technique works for functions that are defined outside of Swift; in the case of `dispatch_async`, the function is defined as a C-language function and can be transparently used in a portable way.

# Running functions on the main thread

Whenever the UI needs to be updated, the update must be run on the main thread. This can be done using the previous pattern to perform updates as they will always be threaded. However, it can be a pain to remember to do this each time it is required.

It is possible to build a Swift function that takes another function and runs it on the main thread automatically. `NSThread.isMainThread` can be used to determine whether the current thread is the UI thread or not; so to run a block of code on the main thread, regardless of whether it's on the main thread or not, the following can be used:

```
func runOnUIThread(fn:()->()) {
  if NSThread.isMainThread() {
    fn()
  } else {
    dispatch_async(dispatch_get_main_queue(), fn)
  }
}
```

This allows code to be submitted to the background thread using:

```
self.runOnUIThread(self.tableView.reloadData)
```

 Due to the lack of parenthesis, the `reloadData` function is not called, but it is passed in as a function pointer. It is dispatched to the correct thread inside the `runOnUIThread` function.

If there is more than one function that needs to be called, an inline block can be created. As this can be passed as a trailing closure to the `runOnUIThread` method, the parenthesis are optional:

```
self.runOnUIThread {
  self.tableView.backgroundColor = UIColor.redColor()
  self.tableView.reloadData()
  self.tableView.backgroundColor = UIColor.greenColor()
}
```

# Parsing JSON

The most popular mechanism to send structured data over a network is to encode it in **JSON**, which stands for **JavaScript Object Notation**. This provides a hierarchical tree data structure, which can store simple numeric, logical, and string-based types, along with array and dictionary representations.

Both Mac OS X and iOS come with a built-in parser for JSON documents, in the `NSJSONSerialization` class. This provides a means to parse a data object and return an `NSDictionary` that contains the key/value pairs of a JSON object, or an `NSArray` to represent JSON arrays. Other literals are parsed and are represented as either `NSNumber` or `NSString` values.

The JSON parser uses `JSONObjectWithData` to create an object from an `NSData` object containing a string. This is typically the format that is returned by network APIs, and it can be created from an existing string using `dataUsingEncoding` with one of the built-in encoding types, such as `NSUTF8StringEncoding`.

A simple JSON array of numbers can be parsed as follows:

```
let array = "[1,2,3]".dataUsingEncoding(NSUTF8StringEncoding)!
let parsed = try? NSJSONSerialization.JSONObjectWithData(
  array, options:.AllowFragments)
```

The return type of this is an optional `AnyObject`. The optionality represents the fact that the data content may not be valid JSON data. This can be cast to an appropriate type using the `as` keyword; if there is a parsing failure, then an error will be thrown.

The `options` can be used to indicate whether the return type should be mutable or not. Mutable data allows the caller to add or delete items after being returned from the parsing function; if not specified, the return value will be immutable. The `NSJSONReadingOptions` options include `MutableContainers` (containing data structures are mutable), `MutableLeaves` (the child leaves are mutable), and `AllowFragments` (allow nonobject, non-array values to be parsed).

The `SampleTable.json` file (referred to in the `viewDidLoad` method) stores an array of entries, with `title` and `content` fields holding text data per entry:

```
[{"title":"Sample Title","content":"Sample Content"}]
```

To parse the JSON file and entries to the table, replace the `default` clause in the `SampleTable` with the following:

```
default:
  let parsed = try? NSJSONSerialization.JSONObjectWithData(
    data!, options:.AllowFragments) as! NSArray
  for entry in parsed {
    self.items +=
      [(entry["title"] as! String,
        entry["content"] as! String)]
  }
```

Running the application will show the **Sample Title** and **Sample Content** entries in the table, which have been loaded and parsed from the book's GitHub repository.

# Handling errors

If there are problems parsing the JSON data then the return type of the `try?` `JSONObjectWithData` function will return a `nil` value. If the type is implicitly unwrapped, then accessing the element will cause an error:

```
do {
 let parsed = try NSJSONSerialization.JSONObjectWithData(data!,
 options:.AllowFragments) {
  // do something with parsed
} catch let error as NSError {
  self.items += [("Error",
   "Cannot parse JSON \(error.localizedDescription)")]
  // show message to user
}
```

The `parsed` value will be of type `AnyObject?` although the `let` block will implicitly unwrap the value, known as *optional binding*. In the previous section, the code was cast to an `NSArray` directly, but if the returned result contains different types (for example, an `NSDictionary` or one of the fragment types `NSNumber` or `NSString`), then attempting to cast to a type that is incompatible with the runtime type will cause a failure.

The type of the object can be tested with `if [object] is [type]`. However, as the next step is usually to cast it to a different class with `as`, a shorthand form `as?` can perform both the test and the cast in one step:

```
if let array = parsed as? NSArray {
  for entry in array {
    // process elements
  }
} else {
  self.items += [("Error", "JSON is not an array")]
}
```

A `switch` statement can be used to check the type of multiple values at the same time. As the values are optional `AnyObject` objects, they need to be converted to a `String` before they can be used in Swift:

```
for entry in array {
  switch (entry["title"], entry["content"]) {
    case (let title as String, let content as String):
      self.items += [(title,content)]
    default:
      self.items += [("Error", "Missing unknown entry")]
  }
}
```

Now when the application is run, any errors are detected and handled without the application crashing.

# Parsing XML

Although JSON is more commonly used, there are still many XML-based network services. Fortunately XML parsing has existed in iOS since version 5 in the `NSXMLParser` class and is simple to access from Swift. For example, some data feeds (such as blog posts) use XML documents, such as Atom or RSS.

The `NSXMLParser` is a stream-oriented parser; that is, it reports individual elements as they are seen. The parser calls the `delegate` to notify when elements are seen and have finished. When an element is seen, the parser also includes any attributes that were present; and for text nodes, the string content. Parsing an XML file involves some state management in the parser. The example used in this section will be to parse an Atom (news feed) file, whose (simplified) structure looks like this:

```
<feed xmlns="http://www.w3.org/2005/Atom">
  <title>AlBlue's Blog</title>
  <link href="http://alblue.bandlem.com/atom.xml" rel="self"/>
```

```
<entry>
    <title type="html">QConLondon and Swift Essentials</title>
    <link href="http://alblue.bandlem.com/2015/01/qcon-swift-
essentials.html"/>
    ...
  </entry>
  ...
</feed>
```

In this case, the goal is to extract all the `entry` elements from the feed, specifically the `title` and the `link`. This presents a few challenges that will become apparent later on.

# Creating a parser delegate

Parsing an XML file requires creating a class that conforms to the `NSXMLParserDelegate` protocol. To do this, create a new class, `FeedParser`, that extends `NSObject` and conforms to the `NSXMLParserDelegate` protocol.

It should have an `init` method that takes an `NSData`, and an `items` property that will be used to acquire the results after they have been parsed:

```
class FeedParser: NSObject, NSXMLParserDelegate {
  var items:[(String,String)] = []
  init(_ data:NSData) {
    // parse XML
  }
}
```

> The `NSXMLParserDelegate` protocol requires that the object also conform to the `NSObjectProtocol`. The easiest way to do this is to subclass `NSObject`. The first mentioned super type is the super class; the second and subsequent super types must be protocols.

# Downloading the data

The XML parser can either parse a stream of data as it is downloaded, or it can take an `NSData` object that has been downloaded previously. On successful download, the `FeedParser` can be used to parse the `NSData` instance and return the list of items.

Although individual expressions can be assigned temporary values that are similar to last time, the statement can be written in a single line (although please note that the error handling is not present). Add the following to the end of the `viewDidLoad` method of `SampleTable`:

```
session.dataTaskWithURL(
  NSURL(string:"https://alblue.bandlem.com/Tag/swift/atom.xml")!,
  completionHandler: {data,response,error -> Void in
    if let data = data {
      self.items += FeedParser(data).items
      self.runOnUIThread(self.tableView.reloadData)
    }
}).resume()
```

This will download the Atom XML feed for the Swift posts from the author's blog at `https://alblue.bandlem.com`. Currently, the data is not parsed, so nothing will be added to the table in this step.

> Make sure that both the download operation and the parsing are handled off the main thread as both of these operations may take some time. Once the data is downloaded, it can be parsed, and after it is parsed, the UI can be notified to redisplay the contents.

# Parsing the data

To process the downloaded XML file, it is necessary to parse the data. This involves writing a parser delegate to listen for the `title` and `link` elements. However, the `title` and `link` elements exist both at the individual `entry` level and also at the top level of the blog. It is therefore necessary to represent some kind of state in the parser, which detects when the parser is inside an `entry` element to allow the correct values to be used.

Elements are reported with the `parser:didStartElement:` method and the `parser:didEndElement:` method. This can be used to determine if the parser is inside an `entry` element by setting a boolean value when an `entry` element starts and resetting it when the `entry` element ends. Add the following to the `FeedParser` class:

```
var inEntry:Bool = false
func parser(parser: NSXMLParser,
  didStartElement elementName: String,
  namespaceURI: String?, qualifiedName:
  String?, attributes: [String:String]) {
  switch elementName {
    case "entry":
```

```
      inEntry = true
    default: break
  }
}
```

The `link` stores the value of the references in an `href` attribute of the element. This is passed when the start element is called, so it is trivial to store. At this point, the title may not be known, so the value of the `link` has to be stored in an optional field:

```
var link:String?
...
// in parser:didStartElement method
case "entry":
  inEntry = true
case "link":
  link = attributes["href"]
default break;
```

The `title` stores its data as a text node, which needs to be implemented with another boolean flag indicating whether the parser is inside a `title` node. Text nodes are reported with the `parser:foundCharacters:` delegate method. Add the following to the `FeedParser`:

```
var title:String?
var inTitle: Bool = false
...
// in parser:didStartElement method
case "entry":
  inEntry = true
case "title":
  inTitle = true
case "link":
...
func parser(parser: NSXMLParser, foundCharacters string:String) {
  if inEntry && inTitle {
    title = string
  }
}
```

By storing the `title` and `link` as optional fields when the end of the `entry` element is seen, the fields can be appended into the `items` list, followed by resetting the state of the parser:

```
func parser(parser: NSXMLParser,
  didEndElement elementName: String,
  namespaceURI: String?, qualifiedName: String?) {
```

```
    switch elementName {
      case "entry":
        inEntry = false
        if title != nil && link != nil {
          items += [(title!,link!)]
        }
        title = nil
        link = nil
      case "title":
        inTitle = false
      default: break
    }
  }
}
```

Finally, having implemented the callback methods, the remaining steps are to create an `NSXMLParser` from the data passed in previously, set the `delegate` (and optionally, the namespace handling), and then invoke the parser:

```
init(_ data:NSData) {
  let parser = NSXMLParser(data: data)
  parser.shouldProcessNamespaces = true
  super.init()
  parser.delegate = self
  parser.parse()
}
```

The assignment of `self` to the `delegate` cannot be done until after `super.init` has been called.

Now when the application is run, a set of news feed items will be displayed.

If running on iOS 9 targets and downloading from http sites, a **App Transport Security has blocked a cleartext HTTP resource load** message may be seen in the console. The solution to fix this is to add an exception in the `Info.plist` file, which permits connections via HTTP, either for the explicit domain or for all domains. Add the following to the `Info.plist` after the first `<dict>` element:

```
<key>NSAppTransportSecurity</key>
<dict>
  <key>NSAllowsArbitraryLoads</key>
  <true/>
</dict>
```

Now when the application is run, the error should no longer be seen.

# Direct network connections

Although most application networking will involve downloading content over standard protocols, such as HTTP(S), and using standard representations, there are times when having a specific data stream protocol is required. In this case, a *stream*-oriented process will allow individual bytes to be read or written, or a *datagram* or *packet*-oriented process can be used to send individual packets of data.

There are networking libraries to support both; an NSStream higher-level Objective-C based class provides a mechanism to drive stream-based responses, and although lower-level packet connections are possible with the CoreFoundation or the POSIX layer, local multiplayer gaming using the MultipeerConnectivity module is often appropriate.

Local networking with the MultipeerConnectivity module involves creating an MCSession, followed by sendData to send NSData objects to connected peers, and using the MCSessionDelegate to receiveData from connected peers. This is often used to synchronize the state of the world, such as the player's current location or health.

# Opening a stream-based connection

A stream is a reliable, ordered sequence of bytes, which is used by most internet protocols. Streams can be created from a network host and port using the NSStream class method getStreamsToHostWithName. This allows an NSInputStream and NSOutputStream to be acquired at the same time.

As this is an existing Objective-C API, the streams are returned via *inout parameters*. In Swift, this translates to the parameters being passed back with an ampersand (&) and declaring the variables as optional.

The input and output streams can then be used to send data asynchronously or synchronously. Asynchronous mechanisms involve scheduling the data processing on the application's run-loop and is covered in the *Asynchronous reading and writing* section. Synchronous mechanisms use read and write to receive or send buffers of data.

Once the streams have been acquired, they need to be *open* to receive or send data. Forgetting this step will result in no networking data being sent.

To simplify acquiring the streams, the following can be created as an extension of the NSStream class. An extension makes a method appear to come from an original class but is implemented externally to that class. Add a StreamExtensions.swift file to the CustomViews project with the following content:

```
extension NSStream {
  class func open(host:String,_ port:Int)
    -> (NSInputStream, NSOutputStream)? {
    var input:NSInputStream?
    var output:NSOutputStream?
    NSStream.getStreamsToHostWithName(
      host, port: port,
      inputStream: &input,
      outputStream: &output)
    guard let i = input, o = output else {
      return nil
    }
    o.open()
    i.open()
    return (i,o)
  }
}
```

A connection to a remote host can be obtained by calling NSStream.
open(host,port), which returns an open pair of input/output streams.

# Synchronous reading and writing

The NSInputStream method read allows bytes to be read from a stream synchronously, while the NSOutputStream method write allows bytes to be written to a stream. These take different types, but the most common approach is to create an array of bytes [UInt8] in Swift as the buffer, and then read into or out of it with an UnsafeMutablePointer (equivalent to an ampersand in C).

The read and write methods both return a number of bytes read/written. This can be negative (in the case of an error), zero, or positive in the case of bytes having been processed. Both calls take a buffer and a maximum length, though it is not guaranteed that the full maximum length will be processed.

Always check the return value of write or read as it is possible that only part of the buffer has been written. Best practice (for synchronous connections) is to wrap the call in a while loop or have some other form of retry in order to ensure that all the data is written.

# Writing data to NSOutputStream

To make it easier to write NSData content to streams, an extension method on NSOuptutStream can be created that performs a full write, based on the size of the data:

```
extension NSOutputStream {
  func writeData(data:NSData) -> Int {
    let size = data.length
    var completed = 0
    while completed < size {
      let wrote = write(UnsafePointer(data.bytes) +
       completed, maxLength:size - completed)
      if wrote < 0 {
        return wrote
      } else {
        completed += wrote
      }
    }
    return completed
  }
}
```

This code takes an NSData and writes it to the underlying stream, returning the number of bytes written (or a negative value if there are problems). The return value of the write method is checked, and if the value is negative, it is returned to the caller directly. Otherwise, the completed counter is incremented with the number of bytes written.

If the number of written bytes reaches the size of the data requested, then the value is returned. Otherwise the loop recurs, this time starting at the point where it left off.

 Although uncommon in Swift, pointer arithmetic is possible by acquiring an UnsafePointer to the data.bytes array, and then incrementing it by the number of bytes already written. The length of the remaining bytes is calculated with size-completed.

# Reading from an NSInputStream

A similar approach can be used to read a full buffer from an NSInputStream by creating a readBytes method that returns an array of bytes of a known size, and a means to convert this to an NSData for easier processing/parsing:

```
extension NSInputStream {
  func readBytes(size:Int) -> [UInt8]? {
    let buffer = Array<UInt8>(count:size,repeatedValue:0)
    var completed = 0
    while completed < size {
      let read = self.read(
      UnsafeMutablePointer(buffer) + completed,
      maxLength: size - completed)
      if read < 0 {
        return nil
      } else {
        completed += read
      }
    }
    return buffer
  }
  func readData(size:Int) -> NSData? {
    if let buffer = readBytes(size) {
      return NSData(
      bytes: UnsafeMutablePointer(buffer),
      length: buffer.count)
    } else {
      return nil
    }
  }
}
```

The readData method returns an NSData, while the readBytes method returns an array of UInt8 values. The NSData approach is useful in some cases (particularly, creating a String from the returned data), and in other cases, being able to process the bytes directly is useful (for example, parsing binary formats). Having both allows either to be used as appropriate.

 Synchronous reads can block forever; if the client application requests exactly 10 bytes but the server only sends 9 bytes, then it will hang permanently until the tenth byte is sent. It is best practice to use asynchronous reads, which cannot block in this way.

# Reading and writing hexadecimal and UTF8 data

Being able to process data as UTF8 values or hexadecimal values can be useful in some protocols. Although both NSString and NSData provide means to convert to and from UTF8, the syntax is overly verbose as it is based on pre-existing Objective-C methods.

To facilitate the conversions, extension methods can be created to provide a simple way of converting to and from UTF8 representations. In addition to class and instance functions, it is possible to use extensions to add dynamic properties to an existing object. This can be used to create utf8data and utf8string properties on NSData and String by adding extensions in a file Extensions.swift, as follows:

```
extension NSData {
  var utf8string:String {
    return String(data:self,
      encoding:NSUTF8StringEncoding)!
  }
}
extension String {
  var utf8data:NSData {
    return self.dataUsingEncoding(
      NSUTF8StringEncoding, allowLossyConversion: false)!
  }
}
```

This allows expressions, such as data.utf8string and string.utf8data, which are much more compact. Each time the expression is evaluated, the associated getter function will be called.

> There is no standard convention to name extensions in Swift at the time this book was written. If there are extensions to a single type of data—such as the streams previously—then the file can be named [Type]Extensions.swift. Alternatively, the name can be used for the type of methods that are called; for example, in this case, UTF8Extensions.swift could have been used.

Parsing hexadecimal data from strings and integers can also be added to the String and Int types, as follows:

```
extension String {
  func fromHex() -> Int {
    var result = 0
    for c in self.characters {
      result *= 16
```

```
          switch c {
          case "0":result += 0      case "1":result += 1
          case "2":result += 2      case "3":result += 3
          case "4":result += 4      case "5":result += 5
          case "6":result += 6      case "7":result += 7
          case "8":result += 8      case "9":result += 9
          case "a","A":result += 10 case "b","B":result += 11
          case "c","C":result += 12 case "d","D":result += 13
          case "e","E":result += 14 case "f","F":result += 15
          default: break
          }
        }
        return result;
      }
  }
  extension Int {
    func toHex(digits:Int) -> String {
      return String(format:"%0\(digits)x",self)
    }
  }
```

This allows hex values to be created with int.toHex and string.fromHex.

# Implementing the Git protocol

It is possible to write a client to query a remote git server using the git:// protocol to determine the hashes of remote tags/branches/references.

 The git:// protocol works by sending *packet lines* of data with each line prefixed with four hexadecimal digits in ASCII, indicating the length of the rest of the data (including the four initial digits). Sending a git-upload-pack request will return a list of references on the remote repository.

As the git:// protocol uses packet lines, create a PacketLineExtensions.swift file with the following content:

```
extension NSOutputStream {
  func writePacketLine(message:String = "") -> Int {
    let data = message.utf8data
    let length = data.length
    if length == 0 {
      return writeData("0000".utf8data)
    } else {
```

```
        let prefix = (length + 4).toHex(4).utf8data
        return self.writeData(prefix) + self.writeData(data)
      }
    }
  }
```

When an empty `NSData` object is passed, the special packet line `0000` is written, indicating the end of the conversation. When a non-empty `NSData` is written, the length of the data is written as a hexadecimal value (including the 4 bytes for the length), followed by the data itself.

This will result in a protocol conversation such as:

```
> 004egit-upload-pack /alblue/com.packtpub.swift.
essentials.git\0host=github.com\0
< 00dfadaa46b98ce211ff819f0bb343395ad6a2ec6ef1
HEAD\0multi_ack thin-pack side-band side-band-
64k ofs-delta shallow no-progress include-tag
multi_ack_detailed symref=HEAD:refs/heads/master
agent=git/2:2.1.1+github-611-gd89bd9f
< 003fadaa46b98ce211ff819f0bb343395ad6a2ec6ef1
refs/heads/master
> 0000
< 0000
```

Reading a packet line is similar:

```
extension NSInputStream {
  func readPacketLine() -> NSData? {
    if let data = readData(4) {
      let length = data.utf8string.fromHex()
      if length == 0 {
        return nil
      } else {
        return readData(length - 4)
      }
    } else {
      return nil
    }
  }
  func readPacketLineString() -> NSString? {
    if let data = self.readPacketLine() {
      return data.utf8string
```

```
        } else {
          return nil
        }
      }
    }
```

In this case, the first 4 bytes are read to determine what the remaining length is. If it is zero, a `nil` value is returned to indicate the end of stream. If it is non-zero, the data is read (less the 4 that is used for the packet line length header). An additional `readPacketLineString` is provided to allow an easy creation of the packet line as an `NSString`.

# Listing git references remotely

To remotely query a git repository for references, the `git-upload-pack` command needs to be sent along with a reference to the repository in question, and optionally, a host. To provide an API to query this programmatically, create a `RemoteGitRepository` class with an initializer that stores the host, port, and repository, and an `lsRemote` function, which returns the value of the references:

```
class RemoteGitRepository {
  let host:String
  let repo:String
  let port:Int
  init(host:String, repo:String, _ port:Int = 9418) {
    self.host = host
    self.repo = repo
    self.port = port
  }
  func lsRemote() -> [String:String] {
    var refs = [String:String]()
    // load the data
    return refs
  }
}
```

To load the data from the repository, a connection to the remote host needs to be made on the default port (in this case, `9418` is the default for the `git://` protocol). Once the streams are opened, the `git-upload-pack` `[repository]\0host=[host]\0` packet line is sent, and subsequently, lines can be read of the form `hash reference`. Add the following to the `lsRemote` function:

```
    // load the data
    if let (input,output) = NSStream.open(host,port) {
      output.writePacketLine(
```

```
      "git-upload-pack \(repo)\0host=\(host)\0")
    while true {
      if let response = input.readPacketLineString() {
        let hash = String(response.substringToIndex(41))
        let ref = String(response.substringFromIndex(41))
        if ref.hasPrefix("HEAD") {
          continue
        } else {
          refs[ref] = hash
        }
      } else {
        break
      }
    }
  }
  output.writePacketLine()
  input.close()
  output.close()
}
```

Calling the `lsRemote` function on a `RemoteGitRepository` instance with an appropriate `host` and `repo` will return a list of hashes by reference.

## Integrating the network call into the UI

As the network can introduce delays or can even result in a complete failure, network calls should never be performed on the UI thread. Previously, the `SampleTable` was used to introduce a `runOnUIThread` function. A similar approach can be used to run a function on a background thread. Add the following to the `SampleTable` class:

```
func runOnBackgroundThread(fn:()->()) {
  dispatch_async(
    dispatch_get_global_queue(
      DISPATCH_QUEUE_PRIORITY_DEFAULT, 0)
    ,fn)
}
```

This will permit `viewDidLoad` to invoke a call in order to query the remote references from the repository, and add them to the table. As before, the call to update the table must be called from the UI thread. Add the following to the end of the `viewDidLoad` method:

```
runOnBackgroundThread {
  let repo = RemoteGitRepository(host: "github.com",
    repo: "/alblue/com.packtpub.swift.essentials.git")
```

```
for (ref,hash) in repo.lsRemote() {
    self.items += [(ref,hash)]
}
self.runOnUIThread(self.tableView.reloadData)
}
```

Now when the application is launched, entries corresponding to the branches and tags in the remote repository should be added to the table.

# Asynchronous reading and writing

As well as synchronous reading and writing, it is also possible to perform *asynchronous* reading and writing. Instead of spinning in a `while` loop, the application can be use callbacks scheduled on the application's run loop.

To receive callbacks, a class that implements `NSStreamDelegate` must be created and assigned to the stream's `delegate` field. When events occur, the `stream` method is called with the type of event and the associated stream.

The stream is registered with `scheduleInRunLoop` (using `NSRunLoop.mainRunLoop()` with a `NSDefaultRunLoopMode` mode). Finally, the stream can be opened.

 If the stream is opened before the delegate is set or scheduled in the run loop, then events will not be delivered.

Events are defined in the `NSStreamEvent` class, and they include `HasSpaceAvailable` (for output streams) and `HasBytesAvailable` (for input streams). By responding to callbacks, the application can process results asynchronously.

 When using Swift, the `NSStreamDelegate` is treated as a weak delegate on the input stream or output stream. This presents problems when using an inline class to provide input parsing; doing so will result in an EXC_ BAD_ACCESS as the delegate is automatically reclaimed by the runtime. This can be avoided by storing a strong circular reference to `self` in the initializer and assigning it to `nil` when the streams are closed.

# Reading data asynchronously from an NSInputStream

This is especially useful for asynchronous protocols, such as XMPP, which may send additional messages at arbitrary times. It also allows battery-powered devices to not spin the CPU if the remote server is slow or hangs.

To receive data asynchronously, a delegate must implement the `NSStreamDelegate` method `stream(stream:handleEvent)`. When data is available, the `HasBytesAvailable` event will be sent, and data can be read accordingly.

To convert the previous example to an asynchronous form, a few changes need to be made. Firstly, the `open` extension method that was created in *Opening a stream connection* section needs to be augmented with a `connect` method, but which does not perform the `open` immediately:

```
class func open(host:String,_ port:Int)
  -> (NSInputStream, NSOutputStream)? {
  if let (input,output) = connect(host,port) {
    input.open()
    output.open()
    return (input,output)
  } else {
    return nil
  }
}
class func connect(host:String,_ port:Int)
  -> (NSInputStream, NSOutputStream)? {
    var input:NSInputStream?
    var output:NSOutputStream?
    NSStream.getStreamsToHostWithName(
      host, port: port,
      inputStream: &input,
      outputStream: &output)
    guard let i = input, o = output else {
      return nil
    }
    return (i,o)
  }
}
```

 In order to receive events asynchronously, the delegate must be set and the stream must be scheduled on a run loop before the stream is opened.

# Creating a stream delegate

To create a stream delegate, create a file called `PacketLineParser.swift` with the following content:

```swift
class PacketLineParser: NSObject, NSStreamDelegate {
  let output:NSOutputStream
  let callback:(NSString)->()
  var capture:PacketLineParser?
  init(_ output:NSOutputStream, _ callback:(NSString) -> ()) {
    self.output = output
    self.callback = callback
    super.init()
    capture = self
  }
  func stream(stream: NSStream, handleEvent: NSStreamEvent) {
    let input = stream as! NSInputStream
    if handleEvent == NSStreamEvent.HasBytesAvailable {
      if let line = input.readPacketLineString() {
        callback(line)
      } else {
        output.writePacketLine()
        input.close()
        output.close()
        capture = nil
      }
    }
  }
}
```

This parser has a callback, which is invoked for each packet line read; when the `HasBytesAvailable` event is sent, the line is read (using the same synchronous mechanism as before) and then passed to the callback. Unlike the synchronous approach, there is no `while` loop here—when data is available, it triggers the parsing of the data.

 As this will be assigned to an input stream delegate (which holds a weak reference), it is necessary to capture a cyclic reference to itself with `capture = self` in order to avoid the runtime from evicting the instance. When the streams are closed, the `capture` will be set to `nil`, which will release the instance.

The `readPacketLine` returns `nil` to indicate either an error or a completed stream; in this case, an empty packet line is sent (to tell the remote server that no further interaction is required), and then both streams are closed.

# Dealing with errors

It is necessary to clean up the streams and remove them from run loops, both when the stream content is successful and when communication errors occur. In addition to the `HasBytesAvailable` event, there are also events that are sent when the stream's end is encountered or an error occurs.

These should be handled in the same way as when the connection comes to a natural end; resources should be tidied, and in particular, the streams should be removed from run loop processing. Finally, the cyclic reference should be removed to permit the `delegate` object to be removed.

The existing `close` code can be moved to its own separate function, and additional cases of the stream ending or errors occurring can perform the same cleanup:

```swift
func stream(stream: NSStream, handleEvent: NSStreamEvent) {
  let input = stream as! NSInputStream
  if handleEvent == NSStreamEvent.HasBytesAvailable {
    if let line = input.readPacketLineString() {
      callback(line)
    } else {
      closeStreams(input,output)
    }
  }
  if handleEvent == NSStreamEvent.EndEncountered
  || handleEvent == NSStreamEvent.ErrorOccurred {
    closeStreams(input,output)
  }
}
func closeStreams(input:NSInputStream,_ output:NSOutputStream) {
  if capture != nil {
    capture = nil
    output.removeFromRunLoop(NSRunLoop.mainRunLoop(),
     forMode: NSDefaultRunLoopMode)
    input.removeFromRunLoop(NSRunLoop.mainRunLoop(),
     forMode: NSDefaultRunLoopMode)
    input.delegate = nil
    output.delegate = nil
    if output.streamStatus != NSStreamStatus.Closed {
      output.writePacketLine()
      output.close()
    }
    if input.streamStatus != NSStreamStatus.Closed {
      input.close()
    }
  }
}
```

# Listing references asynchronously

To provide a list of references asynchronously, the delegate has to be set up with a suitable callback that will parse the returned data. Instead of the method returning a dictionary (which would require synchronous blocking), a callback will be passed, which can be called with references as they are found.

 Please note that there are two separate callbacks: the `PacketLineParser` callback (which reads in network data and returns `NSString` instances on a per-packet-line basis), and the reference parsing callback (which translates the `NSString` into a `(String,String)` tuple).

To start the process, the `git-upload-pack` needs to be sent synchronously after which subsequent responses will be processed asynchronously. This can be done by creating a new method, `lsRemoteAsync`, in the `RemoteGitRepository` class, which takes a callback function for the `(String,String)` tuple:

```
func lsRemoteAsync(fn:(String,String) -> ()) {
  if let (input,output) = NSStream.connect(host,port) {
    input.delegate = PacketLineParser(output) {
    (response:NSString) -> () in
      let hash = String(response.substringToIndex(41))
      let ref = String(response.substringFromIndex(41))
      if !ref.hasPrefix("HEAD") {
        fn(ref,hash)
      }
    }
    input.scheduleInRunLoop(NSRunLoop.mainRunLoop(),
     forMode: NSDefaultRunLoopMode)
    input.open()
    output.open()
    output.writePacketLine(
      "git-upload-pack \(repo)\0host=\(host)\0")
  }
}
```

This creates a connection (but without opening the streams), sets the `delegate`, and schedules the run loop for the input stream, and finally, opens both streams for interaction. Once this is done, the initial `git-upload-pack` message is sent as before. At this point the `lsRemoteAsync` method returns, and subsequent events occur when input data is received from the server.

When a line is received through the `PacketLineParser` callback, it is split into a reference and a hash and then hands the results to the callback passed into the argument in the first place.

 Asynchronous programming often involves many callbacks. Instead of a synchronous program that may look like `A;B;C;`, an asynchronous program often looks like `A(callback:B(callback:C))`. When an input trigger occurs—a network request, user interaction, or timer firing—a sequence of actions can occur via these nested callbacks.

Asynchronous pipelines are generally preferred for battery performance reasons as blocking in a `while` spin loop will waste CPU energy until the condition is satisfied.

# Displaying asynchronous references in the UI

To display the asynchronous data to the screen, the callback must be modified to allow individual elements to update the GUI.

In `SampleTable,` instead of calling `repo.lsRemote` (which performs a synchronous lookup), use `repo.lsRemoteAsync` instead. This requires a callback, which can be used to update the table data and causes the view to reload the contents:

```
// for (ref,hash) in repo.lsRemote() {
//   self.items += [(ref,hash)]
// }
repo.lsRemoteAsync() { (ref:String,hash:String) in
  self.items += [(ref,hash)]
  self.runOnUIThread(self.tableView.reloadData)
}
```

Now when the application is run, the references will be updated asynchronously and the UI will not be blocked by a slow or hung server.

# Writing data asynchronously to an NSOutputStream

Asynchronous sending is not as useful as asynchronous reading unless large uploads are required. If there is a lot of data, then it is unlikely to be written synchronously in a single `write` call. It is better to perform any additional writes asynchronously.

To write data asynchronously requires storing the `completed` count as a variable outside of the function. The `write` method can be used to replace the `while` loop as before by writing a segment of the data on each iteration of the stream method. Although the code isn't needed in this example, code would look something like this:

```
...
self.data = data
// initial write to kick off subsequent events
completed = output.write(UnsafePointer(data.bytes),
 maxLength: data.length
...
var completed:Int
var data:NSData?
func stream(stream: NSStream, handleEvent: NSStreamEvent) {
   let output = stream as! NSOutputStream
   if handleEvent == NSStreamEvent.HasSpaceAvailable
   && data != nil {
     let size = data!.length
     completed += output.write(
      UnsafePointer(data!.bytes) + completed,
      maxLength: size - completed)
     if completed == size {
       completed = 0
       data = nil
     }
   }
}
```

Asynchronous data always starts with a call to synchronously write the data. If not all of the data is written (in other words, `completed` < `size`) then subsequent callbacks will occur on the `NSStreamDelegate`. This can then pick up where the `data` value left off using a similar technique to the synchronous case but without a `while` loop. Instead of the iteration blocking to write the whole data value, the stream call will be called multiple times (in effect replacing each iteration of the `while` loop). On the final run, when `completed == size`, the data is released, and the completion counter is reset.

> The stream callback is called enough times to write all the data. If no data is written, then events are no longer called. New data is only written when an additional value is passed. Care must be taken when writing data from different threads as the data value is processed as an instance variable, and overwriting it may cause data to be lost. The reader is invited to extend the single element data into an array of outstanding data elements so that they can be queued up appropriately.

# Summary

This chapter presented the common techniques that are used to deal with networked data in Swift-based applications with a particular focus on how to maximize battery usage on portable devices using asynchronous techniques to access data.

As most network requests are likely to provide either a JSON or XML-based representation over HTTP(S), the first section of this chapter covered using `NSURLSession` and the asynchronous `dataTask` operations to pull data down from a remote server. The second and third sections then presented how this data can be parsed from either JSON or XML depending on the format required.

The last section presented how to make network connections directly to deal with protocols other than HTTP; and as an example, showed how a remote `git` command can be executed to find out what references are available in a remote git repository. This was presented in two forms: as a synchronous API (to demonstrate the technique of how to work with streams, and to explain the git protocol), followed by its conversion to an asynchronous API, which can be used to minimize CPU cycles and, thus, battery usage, to allow other such translations to be performed in the future.

The next chapter will present how to integrate all of the ideas covered in this book into an iOS application to display GitHub repositories.

# 7
# Building a Repository Browser

Having covered how to integrate the components necessary to build an application, this chapter will create a repository browser that allows user repositories to be displayed using the GitHub API.

This chapter will present the following topics:

- An overview of the GitHub API
- Talking to the GitHub API with Swift
- Creating a repository browser
- Maintaining selection between view controllers

## An overview of the GitHub API

The GitHub API provides a REST-based interface using JSON to return information about users and repositories. Version 3 of the API is documented at `https://developer.github.com/v3/` and is the version used in this book.

The API is rate limited; at the time of writing, anonymous requests can be made up to sixty times per hour, while logged in users have a higher limit. The code repository for this book has sample responses that can be used for testing and development purposes.

# Root endpoint

The main entry point to GitHub is the *root endpoint*. For the main GitHub site, this is `https://api.github.com`, and for GitHub Enterprise installations, it will be of the form `https://hostname.example.org/api/v3/` along with user credentials. The endpoint provides a collection of URLs that can be used to find specific resources:

```
{
  ...
  "issue_search_url": "https://api.github.com/search/issues?q={query}
{&page,per_page,sort,order}",
  "issues_url": "https://api.github.com/issues",
  "repository_url": "https://api.github.com/repos/{owner}/{repo}",
  "user_url": "https://api.github.com/users/{user}"  "user_
repositories_url": "https://api.github.com/users/{user}/
repos{?type,page,per_page,sort}",
}
```

The services are *URI templates*. Text in braces {} is replaced on demand with the values of parameters; text that starts with {?a,b,c} is expanded to form ?a=&b=&c= if present, and is missing otherwise. For example, with a user of alblue, the user_url of the user resource at `https://api.github.com/users/{user}` becomes `https://api.github.com/users/alblue`.

# User resource

The user resource for a specific user contains information about their repositories (repos_url), name, and other information, such as a location and blog (if provided). In addition, the avatar_url provides a URL to an image that can be used to display the user's avatar. For example, `https://api.github.com/users/alblue` contains:

```
{
  ...
  "login": "alblue",
  "avatar_url": "https://avatars.githubusercontent.com/u/76791?v=2",
  "repos_url": "https://api.github.com/users/alblue/repos",
  "name": "Alex Blewitt",
  "blog": "http://alblue.bandlem.com",
  "location": "Milton Keynes, UK",
  ...
}
```

The repos_url link can be used to find the user's repositories. This is what is reported at the root endpoint as the user_repositories_url with the {user} already replaced with the username.

# Repositories resource

Repositories for a user can be accessed via the `repos_url` or `user_repositories_url` references. This returns an array of JSON objects containing information, such as:

```
[{
  "name": "com.packtpub.e4.swift.essentials",
  "html_url":
    "https://github.com/alblue/com.packtpub.swift.essentials",
  "clone_url":
    "https://github.com/alblue/com.packtpub.swift.essentials.git",
  "description": "Swift Essentials",
},{
  "name": "com.packtpub.e4",
  "html_url":
    "https://github.com/alblue/com.packtpub.e4",
  "clone_url":
    "https://github.com/alblue/com.packtpub.e4.git",
  "description":
    "Eclipse Plugin Development by Example: Beginners Guide",
},{
  "name": "com.packtpub.e4.advanced",
  "html_url":
    "https://github.com/alblue/com.packtpub.e4.advanced",
  "clone_url":
    "https://github.com/alblue/com.packtpub.e4.advanced.git",
  "description":
    "Advanced Eclipse plug-in development",
}...]
```

# Repository browser project

The `RepositoryBrowser` client will be created from the **Master Detail** template. This sets up an empty application that can be used on a large device with a split view controller or a navigator view controller on a small device. In addition to this, actions to add entries are also created.

To create a project with tests, ensure that the **Include Unit Tests** option is selected when creating the project:

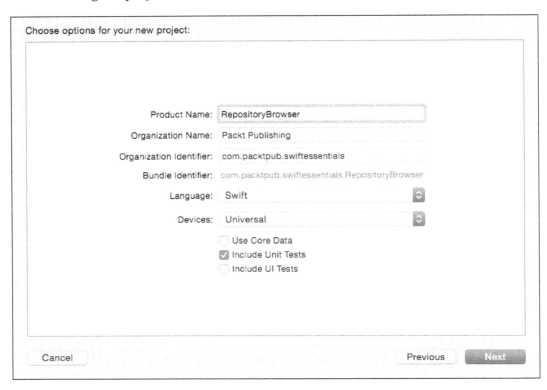

To build the APIs necessary to display content, several utility classes are needed:

- The `URITemplate` class processes URI templates with a set of key/value pairs
- The `Threads` class allows functions to be run in the background or in the main thread
- The `NSURLExtensions` class provides easy parsing of JSON objects from a URL
- The `DictionaryExtensions` class provides a means of creating a Swift dictionary from a JSON object
- The `GitHubAPI` class provides access to the GitHub remote API

# URI templates

URI templates are defined in RFC 6570 at `https://tools.ietf.org/html/rfc6570`. They can be used to replace sequences of text surrounded by {} in a URI. Although GitHub's API uses optional values {?...}, the example client presented in this chapter will not need to use these, and so, they can be ignored in this implementation.

The template class replaces the parameters with values from a dictionary. To create the API, it is useful to write a test case first, following test driven development. A test case class can be created by navigating to **File | New | File... | iOS | Source | Unit Test Case Class** and creating a subclass of XCTestCase in Swift. The test code will look like:

```
import XCTest
class URITemplateTests: XCTestCase {
  func testURITemplate() {
    let template = "http://example.com/{blah}/blah/{?blah}"
    let replacement = URITemplate.replace(
     template,values: ["blah":"foo"])
    XCTAssertEqual("http://example.com/foo/blah/",
     replacement,"Template replacement")
  }
}
```

 Don't forget to ensure that the URITemplateTests.swift file is added to the necessary test targets.

The replace function requires string processing. Although the function can be a class function or an extension on String, having it as a separate class makes testing easier. The function signature looks like:

```
import Foundation
class URITemplate {
  class func replace(template:String, values:[String:String])
   -> String {
   var replacement = template
   while true {
     // replace until no more {…} are present
   }
   return replacement
  }
}
```

Make sure that the URITemplate class is added to the test
target as well; otherwise, the test script will not compile.

The parameters are matched using a regular expression, such as { [^}] }. To search
or access this from a string involves a Range of String.Index values. These are
like integer indexes into the string, but instead of referring to a character by its byte
offset, the index is an abstract representation (some character encodings, such as
UTF8, use multiple bytes to represent a single character).

The rangeOfString method takes a string or regular expression and returns a range
if there is a match present (or nil if there isn't). This can be used to detect whether a
pattern is present or to break out of the while loop:

```
// replace until no more {…} are present
if let parameterRange = replacement.rangeOfString(
  "\\{[^}]*\\}",
  options: NSStringCompareOptions.RegularExpressionSearch) {
  // perform a replacement of parameterRange
} else {
  break
}
```

The parameterRange contains a start and end index that represent the locations
of the { and } characters. The value of the parameter can be extracted with
replacement.substringWithRange(parameterRange). If it starts with {? it is
replaced with an empty string:

```
// perform a replacement of parameterRange
var value:String
let parameter = replacement.substringWithRange(parameterRange)
if parameter.hasPrefix("{?") {
  value = ""
} else {
  // substitute with real replacement
}
replacement.replaceRange(parameterRange, with: value)
```

Finally, if the replacement is of the form {user}, then the value of user is acquired
from the dictionary and used as the replacement value. To get the name of the
parameter, startIndex has to be advanced to the successor, and endIndex has to
be reversed to the predecessor to account for the { and } characters:

```
// substitute with real replacement
let start = parameterRange.startIndex.successor()
```

```
let end = parameterRange.endIndex.predecessor()
let name = replacement.substringWithRange(
 Range<String.Index>(start:start,end:end))
value = values[name] ?? ""
```

Now when the test is run by navigating to **Product** | **Test** or by pressing
*Command + U*, the string replacement will pass.

 The ?? is an optional test that is used to return the first
argument if it is present, and the second argument if it is nil.

# Background threading

Background threading allows functions to be trivially launched on the UI thread
or on a background thread as appropriate. This was explained in *Chapter 6, Parsing
Networked Data*, in the *Networking and user interface* section. Add the following as
Threads.swift:

```
import Foundation
class Threads {
  class func runOnBackgroundThread(fn:()->()) {
    dispatch_async(dispatch_get_global_queue(
      DISPATCH_QUEUE_PRIORITY_DEFAULT, 0),fn)
  }
  class func runOnUIThread(fn:()->()) {
    if NSMainThread.isMainThread() {
      fn()
    } else {
      dispatch_async(dispatch_get_main_queue(), fn)
    }
  }
}
```

The Threads class can be tested with the following test case:

```
import XCTest
class ThreadsTest: XCTestCase {
  func testThreads() {
    Threads.runOnBackgroundThread {
      XCTAssertFalse(NSThread.isMainThread(),
        "Running on background thread")
      Threads.runOnUIThread {
        XCTAssertTrue(NSThread.isMainThread(),
```

```
                "Running on UI thread")
            }
        }
    }
}
```

When the tests are run with *Command + U*, the tests should pass.

# Parsing JSON dictionaries

As many network responses are returned in JSON format and to make JSON parsing easier, extensions can be added to the NSURL class to facilitate the acquiring and parsing of content that is loaded from network locations. Instead of designing a synchronous extension that blocks until data is available, using a callback function is best practice. Create a file NSURLExtensions.swift with the following content:

```
import Foundation
extension NSURL {
  func withJSONDictionary(fn:[String:String] -> ()) {
    let session = NSURLSession.sharedSession()
    session.dataTaskWithURL(self) {
      data,response,error -> () in
      if let json = try? NSJSONSerialization.JSONObjectWithData(
        data!, options: .AllowFragments) as? [String:AnyObject] {
        fn(json!) // will give a compile time error
      } else {
        fn([String:String]())
      }
    }.resume()
  }
}
```

This provides an extension for an NSURL to provide a JSON dictionary. However, the data type returned from the JSONObjectWithData method is [String:AnyObject], not [String:String]. Although it may be expected that it could just be cast to the right type, the as will perform a test, and if there are mixed values (such as a number or a nil), then the entire object will be considered invalid. Instead, the JSON data structure must be converted to a [String:String] type. Add the following as a standalone function to NSURLExtensions.swift:

```
func toStringString(dict:[String:AnyObject]) -> [String:String] {
  var result:[String:String] = [:]
  for (key,value) in dict {
    if let valueString = value as? String {
      result[key] = valueString
```

```
    } else {
      result[key] = "\(value)"
    }
  }
  return result
}
```

This can be used to convert the [String:AnyObject] in the JSON function:

```
fn(toStringString(json!)) // fixes compile time error
```

The function can be tested with a test class using the data: protocol by passing in a *base64* encoded string representing the JSON data. To create a base64 representation, create a string, convert it to a UTF8 data object and then convert it back to a string representation with a data: prefix:

```
import XCTest
class NSURLExtensionsTest: XCTestCase {
  func testNSURLJSON() {
    let json = "{\"test\":\"value\"}".
     dataUsingEncoding(NSUTF8StringEncoding)!
    let base64 = json.base64EncodedDataWithOptions(
.EncodingEndLineWithLineFeed)
    let data = String(data: base64,
     encoding: NSUTF8StringEncoding)!
    let dataURL = NSURL(string:"data:text/plain;base64,\(data)")!
    dataURL.withJSONDictionary {
      dict in
      XCTAssertEqual(dict["test"] ?? "", "value",
       "Value is as expected")
    }
    sleep(1)
  }
}
```

Please note that the sleep(1) is required as parsing the response has to happen in the background thread and, therefore, may not be immediately available. By adding a delay to the function it gives a chance for the assertion to be executed.

# Parsing JSON arrays of dictionaries

A similar approach can be used to parse arrays of dictionaries (such as those that are returned by the list repositories resource). The differences here are the type signatures (which have an extra [] to represent the array), and the fact that a map is being used to process the elements in the list:

```
func withJSONArrayOfDictionary(fn:[[String:String]] -> ()) {
  ...
  if let json = try? NSJSONSerialization.JSONObjectWithData(
   data, options: .AllowFragments) as? [[String:AnyObject]] {
    fn(json!.map(toStringString))
  } else {
    fn([[String:String]]())
  }
}
```

The test can be extended as well:

```
let json = "[{\"test\":\"value\"}]".
 dataUsingEncoding(NSUTF8StringEncoding)!
...
dataURL.withJSONArrayOfDictionary {
  dict in XCTAssertEqual(dict[0]["test"] ?? "", "value",
 "Value is as expected")
}
```

# Creating the client

Now that the utilities are complete, the GitHub client API can be created. Once that is complete, it can be integrated with the user interface.

# Talking to the GitHub API

A Swift class will be created to talk to the GitHub API. This will connect to the root endpoint host and download the JSON for the service URLs so that subsequent network connections can be made.

To ensure that network requests are not repeated, an NSCache will be used to save the responses. This will automatically be emptied when the application is under memory pressure:

```
import Foundation
class GitHubAPI {
  let base:NSURL
```

```
    let services:[String:String]
    let cache = NSCache()
    class func connect() -> GitHubAPI? {
      return connect("https://api.github.com")
    }
    class func connect(url:String) -> GitHubAPI? {
      if let nsurl = NSURL(string:url) {
        return connect(nsurl)
      } else {
        return nil
      }
    }
    class func connect(url:NSURL) -> GitHubAPI? {
      if let data = NSData(contentsOfURL:url) {
        if let json = try? NSJSONSerialization.JSONObjectWithData(
         data,options:.AllowFragments) as? [String:String] {
          return GitHubAPI(url,json!)
        } else {
         return nil
        }
      } else {
        return nil
      }
    }
    init(_ base:NSURL, _ services:[String:String]) {
      self.base = base
      self.services = services
    }
  }
```

This can be tested by saving the response from the main GitHub API site at `https://api.github.com` into an `api/index.json` file by creating an `api` directory in the root level of the project and running `curl https://api.github.com > api/index.json` from a Terminal prompt. Inside Xcode, add the `api` directory to the project by navigating to **File | Add Files to Project...** or by pressing *Command + Option + A*, and ensure it is associated with the test target.

It can then be accessed with an `NSBundle`:

```
import XCTest
class GitHubAPITests: XCTestCase{
  func testApi() {
    let bundle = NSBundle(forClass:GitHubAPITests.self)
    if let url = bundle.URLForResource("api/index",
     withExtension:"json") {
```

```
      if let api = GitHubAPI.connect(url) {
        XCTAssertTrue(true,"Created API \(api)")
      } else {
        XCTAssertFalse(true,"Failed to parse \(url)")
      }
    } else {
      XCTAssertFalse(true,"Failed to find sample API")
    }
  }
}
```

 The dummy API should not be part of the main application's target, but rather of the test target. As a result, instead of using `NSBundle.mainBundle` to acquire the application's bundle, `NSBundle(forClass)` is used.

# Returning repositories for a user

The APIs returned from the services lookup include `user_repositories_url`, which is a template that can be instantiated with a specific user. It is possible to add a method `getURLForUserRepos` to the `GitHubAPI` class that will return the URL of the user's repositories. As it will be called frequently, the results should be cached using an `NSCache`:

```
func getURLForUserRepos(user:String) -> NSURL {
  let key = "r:\(user)"
  if let url = cache.objectForKey(key) as? NSURL {
    return url
  } else {
    let userRepositoriesURL = services["user_repositories_url"]!
    let userRepositoryURL = URITemplate.replace(
     userRepositoriesURL, values:["user":user])
    let url = NSURL(string:userRepositoryURL, relativeToURL:base)!
    cache.setObject(url, forKey:key)
    return url
  }
}
```

Once the URL is known, data can be parsed as an array of JSON objects using an asynchronous callback function to notify when the data is ready:

```
func withUserRepos(user:String, fn:([[String:String]]) -> ()) {
  let key = "repos:\(user)"
  if let repos = cache.objectForKey(key) as? [[String:String]] {
```

```
      fn(repos)
    } else {
      let url = getURLForUserRepos(user)
      url.withJSONArrayOfDictionary {
        repos in
        self.cache.setObject(repos,forKey:key)
        fn(repos)
      }
    }
}
```

This can be tested using a simple addition to the `GitHubAPITests` class:

```
api.withUserRepos("alblue") {
  array in
  XCTAssertEqual(24,array.count,"Number of repos")
}
```

> The sample data contains 24 repositories in the following file, but the
> GitHub API may contain a different value for this user in the future:
> https://raw.githubusercontent.com/alblue/com.
> packtpub.swift.essentials/master/RepositoryBrowser/
> api/users/alblue/repos.json

# Accessing data through the AppDelegate

When building an iOS application that manages data, deciding where to declare
the variable is the first decision that has to be made. When implementing a view
controller, it is common for view-specific data to be associated with that class; but if
the data needs to be used across multiple view controllers, there is more choice.

A common approach is to wrap everything into a *singleton*, which is an object
that is instantiated once. This is typically achieved with a `private var` in the
implementation class, with a `class func` that returns (or instantiates on demand)
the singleton.

> The Swift `private` keyword ensures that the variable is only visible
> in the current source file. The default visibility is `internal`, which
> means that code is only visible in the current module; the `public`
> keyword means that it is visible outside of the module as well.

Another approach is to use the `AppDelegate` itself. This is in effect already a singleton that can be accessed with `UIApplication.sharedApplication().delegate`, and is set up prior to any other object accessing it.

 The AppDelegate should not be overused to store data. Instead of adding too many properties, consider creating a separate class or struct to hold the values.

The `AppDelegate` will be used to store the reference to the `GitHubAPI`, which could use a preference store or other external means to define what instance to connect to, along with the list of users and a cache of repositories:

```
class AppDelegate {
  var api:GitHubAPI!
  var users:[String] = []
  var repos:[String:[[String:String]]] = [:]
  func application(application: UIApplication,
    didFinishLaunchingWithOptions: [NSObject: AnyObject]?)
    -> Bool {
    api = GitHubAPI.connect()
    users = ["alblue"]
    return true
  }
}
```

To facilitate loading repositories from view controllers, a function can be added to `AppDelegate` to provide a list of repositories for a given user:

```
func loadRepoNamesFor(user:String, fn:([[String:String]])->()) {
  repos[user] = []
  api.withUserRepos(user) {
    results in
    self.repos[user] = results
    fn(results)
  }
}
```

# Accessing repositories from view controllers

In the `MasterViewController` (created from the **Master Detail** template or a new subclass of a `UITableViewController`), define an instance variable, `AppDelegate`, which is assigned in the `viewDidLoad` method:

```
class MasterViewController:UITableViewController {
  var app:AppDelegate!
  override func viewDidLoad() {
    app = UIApplication.sharedApplication().delegate
     as? AppDelegate
    …
  }
}
```

The table view controller provides data in a number of sections and rows. The `numberOfSections` method will return the number of users with the section title being the username (indexed by the users list):

```
override func numberOfSectionsInTableView(tableView: UITableView)
 -> Int {
  return app.users.count
}
override func tableView(tableView: UITableView,
 titleForHeaderInSection section: Int) -> String? {
  return app.users[section]
}
```

The `numberOfRowsInSection` function is called to determine how many rows are present in each section. If the number is not known, `0` can be returned while running a background query to find the right answer:

```
override func tableView(tableView: UITableView,
 numberOfRowsInSection section: Int) -> Int {
  let user = app.users[section]
  if let repos = app.repos[user] {
    return repos.count
  } else {
    app.loadRepoNamesFor(user) { _ in
      Threads.runOnUIThread {
        tableView.reloadSections(
          NSIndexSet(index: section),
          withRowAnimation: .Automatic)
      }
```

```
        }
    return 0
    }
}
```

 Remember to reload the section on the UI thread; otherwise, the updates won't display correctly.

Finally, the repository name needs to be shown in the value of the cell. If a default UITableViewCell is used, then the value can be set on the textLabel; if it is loaded from a storyboard prototype cell, then the content can be accessed appropriately using tags:

```
override func tableView(tableView: UITableView,
  cellForRowAtIndexPath indexPath: NSIndexPath)
  -> UITableViewCell {
  let cell = tableView.dequeueReusableCellWithIdentifier(
    "Cell", forIndexPath: indexPath)
  let user = app.users[indexPath.section]
  let repo = app.repos[user]![indexPath.row]
  cell.textLabel!.text = repo["name"] ?? ""
  return cell
}
```

When the application is run, the list of repositories will be displayed, grouped by the user:

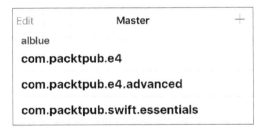

# Adding users

At this moment, the list of users is hardcoded into the application. It would be preferable to remove this hardcoded list and allow users to be added on demand. Create an `addUser` function in the `AppDelegate` class:

```
func addUser(user:String) {
  users += [user]
  users.sortInPlace({ $0 < $1 })
}
```

This allows the detail controller to call the `addUser` function and ensure that the list of users is ordered alphabetically.

 The `$0` and `$1` are anonymous parameters expected by the `sort` function. This is a shorthand form of `users.sort({ user1, user2 in user1 < user2})`. It is also possible to sort the array using the `<` function on the array itself using `users.sortInPlace(<)`.

The add button can be created in the `MasterViewController` in the `viewDidLoad` method such that the `insertNewObject` method is called when tapped:

```
override func viewDidLoad() {
  super.viewDidLoad()
  let addButton = UIBarButtonItem(barButtonSystemItem: .Add,
   target: self, action: "insertNewObject:")
  self.navigationItem.rightBarButtonItem = addButton

  ...
}
```

When the add button is selected, a `UIAlertController` dialog can be shown with a number of actions with handlers that will be called to add the user.

Add (or replace) the `insertNewObject` in the `MasterViewController`, as follows:

```
func insertNewObject(sender: AnyObject) {
  let alert = UIAlertController(
    title: "Add user",
    message: "Please select a user to add",
    preferredStyle: .Alert)
  alert.addAction(UIAlertAction(
    title: "Cancel", style: .Cancel, handler: nil))
  alert.addAction(UIAlertAction(
    title: "Add", style: .Default) {
    alertAction in
```

```
    let username = alert.textFields![0].text
    self.app.addUser(username!)
    Threads.runOnUIThread {
      self.tableView.reloadData()
    }
  })
  alert.addTextFieldWithConfigurationHandler {
    textField -> Void in
    textField.placeholder = "Username";
  }
  presentViewController(alert, animated: true, completion: nil)
}
```

Now, the users can be added in the UI by clicking the **Add** (+) button at the top right of the application. Each time the application is launched, the users array will be empty, and users can be re-added.

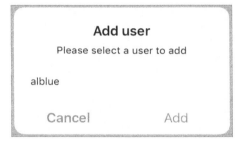

 Users could persist between launches using
NSUserDefaults.standardUserDefaults and the
setObject:forKey and stringArrayForKey methods.
The implementation of this is left to the reader.

# Implementing the detail view

The final step is to implement the detail view so that when a repository is selected, per-repository information is shown. At the time the repository is selected from the master screen, the username, and repository name are known. These can be used to pull more information from the repository and add the items into the detail view.

Update the view in the storyboard to add four labels and four label titles for username, repository name, number of watchers, and number of open issues. Wire these into outlets into the `DetailViewController`:

```
@IBOutlet weak var userLabel: UILabel?
@IBOutlet weak var repoLabel: UILabel?
@IBOutlet weak var issuesLabel: UILabel?
@IBOutlet weak var watchersLabel: UILabel?
```

To set content on the details view, the `user` and `repo` will be stored as (optional) strings, and the additional `data` will be stored in string key/value pairs. When they are changed, the `configureView` method should be called to redisplay content:

```
var user: String? { didSet { configureView() } }
var repo: String? { didSet { configureView() } }
var data:[String:String]? { didSet { configureView() } }
```

The `configureView` call will also need to be called after the `viewDidLoad` method is called to ensure that the UI is set up as expected:

```
override func viewDidLoad() { configureView() }
```

In the `configureView` method, the labels may not have been set, so they need to be tested with an `if let` statement before the content is set:

```
func configureView() {
  if let label = userLabel { label.text = user }
  if let label = repoLabel { label.text = repo }
  if let label = issuesLabel {
    label.text = self.data?["open_issues_count"]
  }
  if let label = watchersLabel {
    label.text = self.data?["watchers_count"]
  }
}
```

If using the standard template, the `splitViewController` of the `AppDelegate` needs to be changed to return `true` after the detail view is amended:

```
func splitViewController(
 splitViewController: UISplitViewController,
 collapseSecondaryViewController
  secondaryViewController:UIViewController!,
 ontoPrimaryViewController
  primaryViewController:UIViewController!) -> Bool {
  return true
}
```

The splitViewController:collapseSecondaryViewController method determines whether or not the first page that is displayed is the master (true) or detail (false) page.

# Transitioning between the master and detail views

The connection between the master and detail view is triggered with the showDetail segue in MasterViewController. This can be used to extract the selected row from the table, which can then be used to extract the selected row and section:

```
override func prepareForSegue(segue: UIStoryboardSegue,
  sender: AnyObject?) {
  if segue.identifier == "showDetail" {
    if let indexPath = self.tableView.indexPathForSelectedRow {
      // get the details controller
      // set the details
    }
  }
}
```

The details controller can be accessed from the segue's destination controller — except that the destination is the navigation controller, so it needs to be unpacked one step further:

```
// get the details controller
let controller = (segue.destinationViewController as!
 UINavigationController).topViewController
 as! DetailViewController
// set the details
```

Next, the details need to be passed in, which can be extracted from indexPath, as in the prior parts of the application:

```
let user = app.users[indexPath.section]
let repo = app.repos[user]![indexPath.row]
controller.repo = repo["name"] ?? ""
controller.user = user
controller.data = repo
```

Finally, to ensure that the application works in split mode with
`SplitViewController`, the back button needs to be displayed if in split mode:

```
controller.navigationItem.leftBarButtonItem =
  self.splitViewController?.displayModeButtonItem()
controller.navigationItem.leftItemsSupplementBackButton = true
```

Running the application now will show a set of repositories, and when one is
selected, the details will be displayed:

| ‹ Master | **Detail** |
|---|---|
| User | alblue |
| Repo | com.packtpub.swift.essentials |
| Issues | 0 |
| Watchers | 13 |

 If a crash is seen when displaying the detail view, check in the
Main.storyboard that the connector for a nonexistent field is
not defined. Otherwise an error similar to **This class is not key
value coding-compliant for the key detailDescriptionLabel**
might be seen, which is caused by the Storyboard runtime
attempting to assign a missing outlet in the code. Open the
Main.storyboard, go to the connections inspector, and
remove the connection to the missing outlet.

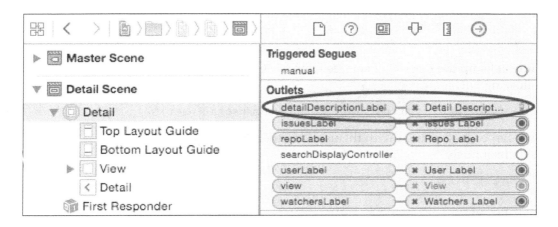

# Loading the user's avatar

The user may have an avatar or icon that they have uploaded to GitHub. This information is stored in the user information, which is accessible from a separate lookup in the GitHub API. Each user's avatar will be stored as a reference with `avatar_url` in the user information document, such as `https://api.github.com/users/alblue`, which will return something like this:

```
{
  ...
  "avatar_url": "https://avatars.githubusercontent.com/u/76791?v=2",
  ...
}
```

This URL represents an image that can be used in the header for the user's repository.

To add support for this, the user info needs to be added to the `GitHubAPI` class:

```
func getURLForUserInfo(user:String) -> NSURL {
  let key = "ui:\(user)"
  if let url = cache.objectForKey(key) as? NSURL {
    return url
  } else {
    let userURL = services["user_url"]!
    let userSpecificURL = URITemplate.replace(userURL,
     values:["user":user])
    let url = NSURL(string:userSpecificURL, relativeToURL:base)!
    cache.setObject(url,forKey:key)
    return url
  }
}
```

This looks up the `user_url` service from the GitHub API, which returns the following URI template:

```
"user_url": "https://api.github.com/users/{user}",
```

This can be instantiated with the user and then the image can be loaded asynchronously:

```
import UIKit
...
func withUserImage(user:String, fn:(UIImage -> ())) {
  let key = "image:\(user)"
  if let image = cache.objectForKey(key) as? UIImage {
    fn(image)
  } else {
```

```
    let url = getURLForUserInfo(user)
    url.withJSONDictionary {
      userInfo in
      if let avatar_url = userInfo["avatar_url"] {
        if let avatarURL = NSURL(string:avatar_url,
          relativeToURL:url) {
          if let data = NSData(contentsOfURL:avatarURL) {
            if let image = UIImage(data: data) {
              self.cache.setObject(image,forKey:key)
              fn(image)
} } } } } } }
```

Once the support to load the user's avatar has been implemented, it can be added to the view's header to display it in the user interface.

 The set of nested `if` statements here suggests that it may be better to refactor to Swift's `guard` statement instead. This would ensure that the indentation does not increase on each condition. The refactoring is left as an exercise for the reader.

# Displaying the user's avatar

The table view that presents the repository information by user can be amended so that along with the user's name, it also displays their avatar at the same time. Currently, this is done in the `tableView:`**`title`**`ForHeaderInSection` method, but an equivalent `tableView:`**`view`**`ForHeaderInSection` method is available that provides more customization options.

Although the method signature indicates that the return type is `UIView`, in fact, it must be a subtype of `UITableViewHeaderFooterView`. Unfortunately, there is no support to edit or customize these in Storyboard, so they must be implemented programmatically.

To implement the `viewForHeaderInSection` method, obtain the username as before, and set it to the `textLabel` of a newly created `UITableViewHeaderFooterView`. Then, in the asynchronous image loader, create a frame that has the same origin but a square size for the image, and then create and add the image as a subview of the header view. The method will look like this:

```
override func tableView(tableView: UITableView,
  viewForHeaderInSection section: Int) -> UIView? {
  let cell = UITableViewHeaderFooterView()
  let user = app.users[section]
  cell.textLabel!.text = user
```

```
app.api.withUserImage(user) {
  image in
  let minSize = min(cell.frame.height, cell.frame.width)
  let squareSize = CGSize(width:minSize, height:minSize)
  let imageFrame = CGRect(origin:cell.frame.origin,
   size:squareSize)
  Threads.runOnUIThread {
    let imageView = UIImageView(image:image)
    imageView.frame = imageFrame
    cell.addSubview(imageView)
    cell.setNeedsLayout()
    cell.setNeedsDisplay()
  }
}
return cell
}
```

Now when the application is run, the avatar will be displayed overlaying the user's repositories:

# Summary

This chapter has shown how to integrate the subjects that were created in this book to integrate them into a functional application to interact with a remote network service, such as GitHub, and be able to present this information in a tabular way.

By ensuring that all network requests are implemented on background threads, and that returned data is updated on the UI thread, the application will remain responsive to the user's input. Graphics and custom views can be created to provide headings, or the Storyboard could be modified to include more graphics for each repository.

# 8
# Adding Watch Support

Apple released watchOS to the public with the release of the Apple Watch in April 2015. However, with the release of watchOS 2 in September 2015, developers have been able to write extensions that run on the watch itself rather than relying on a companion iOS device being available. This chapter will show how to add watch support to the existing Repository Browser application (created in *Chapter 7, Building a Repository Browser*.)

This chapter will present the following topics:

- Adding a watch extension to an existing project
- The type of watch interfaces
- Using tables, text, and images
- How to transition between screens with selected context
- Best practices for watch applications

## Watch applications

A watch application consists of code that can execute on the watch itself. A watch application is developed in Swift and run as a *watch extension* and a *watch app*. For watchOS 2, both run on the watch. (On watchOS 1, the watch extension ran on the companion iPhone.) This chapter will assume watchOS 2 is being used in order to run Swift-compiled code directly on the watch.

As the first version of watchOS did not allow code to be executed on the watch, the code was bundled up into a watch extension, which ran as part of the companion application on the iPhone. The watch app contained resources and other images which were presented directly on the watch. With watchOS 2, the separation became less relevant. A future version of Xcode or watchOS may result in the two concepts becoming combined.

# Adding a watch target

To add watch support for an existing application, a new target must be created for the watch. Open the existing **Repository Browser** application, navigate to **File** | **New** | **Target**, and select **WatchKit App** from the **watchOS** section:

Once this is created, it will ask for the name of the watch application. This can't be the same name as the enclosing project, so call it `RepositoryBrowserWatch` instead. The language should be **Swift**; the other user interface elements (**Complications**, **Glance**, and **Notifications**) are not relevant to this project, and so, it can be deselected:

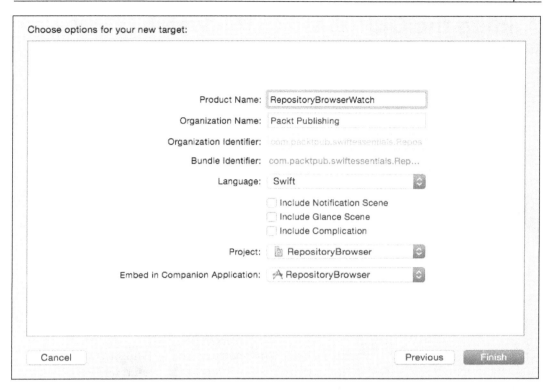

When **Finish** is pressed, the following new elements will be created in the project:

- `RepositoryBrowserWatch`: This is the watch application, which provides the interface descriptions for the application

- `RepositoryBrowserWatch Extension`: This is the content corresponding to the watch application's executable code

- `InterfaceController.swift`: This is the Swift file corresponding to the user interface element that gets automatically created

- `ExtensionDelegate.swift`: This is the Swift file corresponding to the user application as a whole (similar to an `AppDelegate` on a traditional iOS application)

# Adding the GitHubAPI to the watch target

In order to allow the watch application to use the GitHubAPI that was developed in *Chapter 7, Building a Repository Browser*, the following code should be added to the ExtensionDelegate:

```
var api:GitHubAPI!
var users:[String] = []
var repos:[String:[[String:String]]] = [:]
func loadReposFor(user:String, fn:([[String:String]])->()) {
  repos[user] = []
  api.withUserRepos(user) {
    results in
    self.repos[user] = results
    fn(results)
  }
}
func addUser(user:String) {
  users += [user]
  users.sortInPlace({ $0 < $1 })
}
```

This will initially generate a compile-time error because the GitHubAPI class (and the dependent classes) is not currently associated to the watch target. To resolve this, select the GitHubAPI, Threads, NSURLExtensions, and URITemplate Swift files and open the file inspector by pressing *Command + Option + 1* or by navigating to **View | Utilities | Show File Inspector**. Ensure these are added to the **RepositoryBrowserWatch Extension** target by selecting the appropriate checkbox:

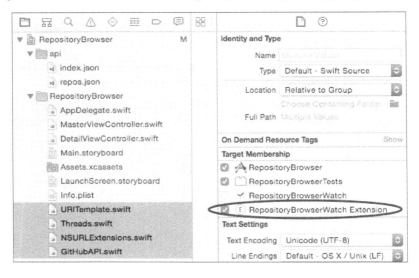

Now when the watch target is built and run, a watch simulator will show up with a black screen and the time at the top-right of the application. If this is not displayed, verify that the target selected is for the watch application:

# Creating watch interfaces

A watch's user interface is built up of elements in a similar way to iOS applications, except that the user toolkit is built using WatchKit instead of UIKit. In the same way that classes, such as UITableView, exist, corresponding classes, such as WKInterfaceTable, also exist. There are minor differences; for example, the UITableView will dynamically populate the elements upon display, but the WKInterfaceTable will expect to be told in advance how many rows exist and what these rows are.

# Adding a list of users to the watch

Unlike the UITableView, which provides section headers to group rows, a WKInterfaceTable only permits a single list of items. Instead, the application will be designed so that the first screen will show a list of users, and then the second screen will show the selected user's repositories.

For testing purposes, add the following into the applicationDidFinishLaunching method of the ExtensionDelegate class:

```
api = GitHubAPI.connect()
addUser("alblue")
```

This will allow other classes to query the `ExtensionDelegate` property `users` to show some content. As with the `AppDelegate` of an iOS application, there is a global singleton that can be accessed. Add the following to the `InterfaceController`:

```
let delegate = WKExtension.sharedExtension().delegate as!
ExtensionDelegate
```

To display a list of users, the interface itself must have a table. Each table row has its own controller class, which can be a simple `NSObject` subclass. To display a list of user names, create a `UserRowController` class that has a single label. As this is a private implementation detail of the `InterfaceController`, it makes sense to include it in the same file:

```
class UserRowController: NSObject {
  @IBOutlet weak var name: WKInterfaceLabel!
}
```

Add the following to the `InterfaceController` class, which will be connected to the interface later:

```
@IBOutlet weak var usersTable: WKInterfaceTable!
```

Now, the table can be populated in the `awakeWithContext` method. This involves setting the number of rows, and the type of the rows. Add the following:

```
let users = delegate.users
usersTable.setNumberOfRows(users.count, withRowType: "user")
for (index,user) in users.enumerate() {
  let controller = usersTable.rowControllerAtIndex(index) as!
UserRowController
  controller.name.setText(user)
}
```

If the application is run at this point, several errors will occur because the `IBOutlet` references have not been connected, and the row type user has not been associated with the `UserRowController` class.

# Wiring up the interface

Having generated the content for the users, the interface must be wired to the implementation detail. Open `Interface.storyboard` in the `RepositoryBrowserWatch` folder and go to **Interface Controller Scene**. This will present a black watch surrounded with a clock and **Any Screen Size** displayed at the bottom. Like iOS application interfaces, they can come in different sizes (38mm or 42mm at the time of writing).

Open the object library by pressing *Command + Option + Control + 3* or by navigating to **View | Utilities | Show Object Library**. Type `table` into the search field and then drag it into the watch interface:

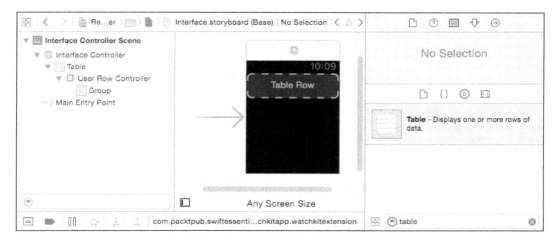

From the **Interface Controller** in the document outline on the left, press *Control* and drag down to the table to create a connection to the `usersTable` outlet that is defined in the interface controller:

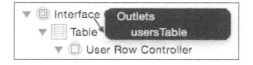

When the `InterfaceController` is instantiated, the `usersTable` will be wired up to the outlet. However, there are still no connections to the rows. To do this, drag a label into the dotted area with the **Table Row** placeholder. To ensure that the label takes up all the available space, set the size to **Relative to Container** with a factor of **1** for both **Width** and **Height**:

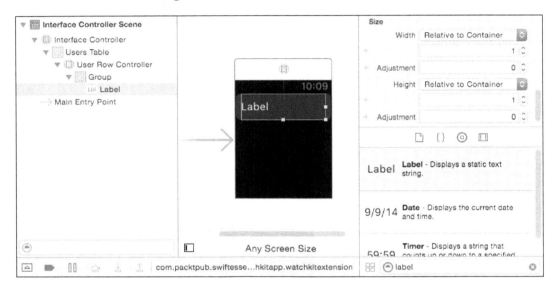

In order to connect the label's text with the `UserRowController`, two things have to be done. Firstly, the type of the row must be set to correspond to the `UserRowController` class, which will allow the label to be wired up to the name outlet. Secondly, the row must be given the identifier user to allow it to be connected with the `rowType` that was specified in the previous section.

To set the row controller's class, open the **Identity Inspector** by pressing the *Command + Option + 3* keys or by navigating to **View | Utilities | Show Identity Inspector**. Choose **UserRowController** from the dropdown, which should also set the module name **RepositoryBrowserWatch_Extension**. Once this is done, the user controller can make a connection to the label by pressing *Control* and dragging to the label, followed by choosing the **name** outlet:

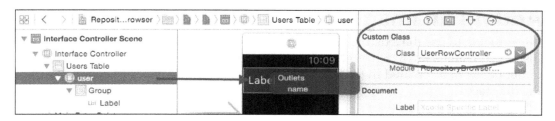

To set the row controller's type, switch to the **Attributes Inspector** by pressing the *Command + Option + 4* keys or by navigating to **View | Utilities | Show Attributes Inspector**, and entering the `rowType` that was used previously, which is `user`:

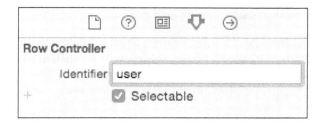

Now when the application is run the list of users should be seen, which includes `alblue`:

# Adding an image

It is possible to use the existing API to return an image for the user, and this can be displayed using a WKInterfaceImage in a similar way to the text name. First, an outlet needs to be created in the UserRowController so that it can be connected to the interface:

```
class UserRowController: NSObject {
  @IBOutlet weak var name: WKInterfaceLabel!
  @IBOutlet weak var icon: WKInterfaceImage!
}
```

The interface now needs to be updated to add the image. This can be done by searching for **image** in the object library and then dragging it into the user row.

The watch prefers that image sizes are known in advance, so the size of the image can be fixed with a size of 32 by 32 pixels, which will be sufficient for both the larger and smaller watch sizes. Marking the image as **Aspect Fit** will ensure that the image doesn't get resized inappropriately, and that the whole image will be displayed.

 It is possible to click the + icon next to the **size** and then specify different dimensions for the two different watches.

Aligning the image on the right and on the center will give the same impression for both sizes of watch. Changing the alignment to **Right** and **Center** will allow the display to adjust to different sizes. It may also make sense to modify the user's name width from **Relative to Container** to **Size to Fit**, but this is not strictly necessary. Finally, connect the outlet from the **user** row with the image using *Control* and dragging the mouse, followed by choosing the **icon** outlet. The resulting user interface will look like this:

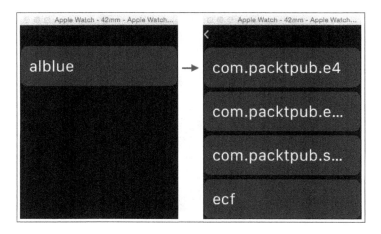

Having created and wired up the image, the last step is to populate the data. In the `InterfaceController` method `awakeFromContext`, after setting the user's name, add a call to the API to acquire the image similar to the `DetailViewController` in the last chapter:

```
controller.name.setText(user) // from before
delegate.api.withUserImage(user) {
  image in controller.icon.setImage(image)
}
```

Now when the application is run, after a brief pause, the user's avatar will be seen:

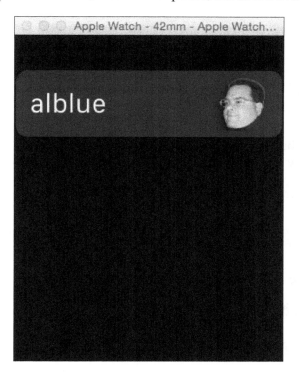

# Responding to user interaction

Typically, a watch user interface will present information to the user or let them select or manipulate it in some way. When items are presented in a table, then it is natural to let the user tap on the row to show a subsequent screen. Watch applications use **segues** to move from one screen to another in a similar way to iOS applications.

The first step will involve creating a new controller file called
`RepositoryListController.swift`. This will be used to hold the
`RepositoryListController` and `RepositoryRowController` classes, in a very
similar way to the existing `InterfaceController`. As with the other view, there will
be a table to store the rows, and each row will have a `name` label:

```
class RepositoryRowController: NSObject {
  @IBOutlet weak var name: WKInterfaceLabel!
}
class RepositoryListController: WKInterfaceController {
  let delegate = WKExtension.sharedExtension().delegate as!
ExtensionDelegate
  @IBOutlet weak var repositoriesTable: WKInterfaceTable!
}
```

 Don't forget to add the `RepositoryListController.swift` file
to the `RepositoryBrowserWatch Extension` target, or it will
not be possible to use that as the implementation class.

Once these classes have been created, the `interface.storyboard` can be opened
and a new **Interface Controller** dragged in from the object library. This will create an
empty screen, which can have other objects added.

 Ensure that the **Interface Controller** is selected, instead of
the **Glance Interface Controller** or the **Notification Interface
Controller** as these are used for different purposes.

Once the interface controller has been created, drag a **Table** from the object library
onto the interface controller, and then drag a **Label** from the object library into the
row placeholder in the same way as in the previous interface controller example.

The interface controller will need to be updated to point to the
`RepositoryListController` class; this can be done by selecting the
interface controller and going to the **Identity Inspector** as before. Once the
`RepositoryListController` implementation is defined, press *Control* and drag it
from the interface controller icon to the table and wire it to the `repositoriesTable`.

 These connections are made in the same way as they
were for the `usersTable` in the previous section.

The row placeholder's class can be defined by selecting the placeholder under the **Repositories Table** in the document outline, and then setting the row controller's identity to `repository` in the **Attributes Inspector**. This will allow the repository row placeholder to connect the name attribute to the label in the scene.

The last connection is to add a **segue** from the users screen to the repositories screen. Press *Control* and drag from the **user** row in the **Users Table** to the repository list controller, and in the popup, select a **Push Segue**.

The final connection will look like this:

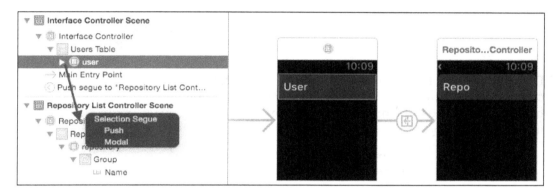

When the user is selected in the first screen, the second screen should slide over. At the moment this will be empty but the repositories will be populated in the next section.

# Adding context and showing repositories

To pass data from one screen to another requires a context to be set. Each `WKInterface` screen has an `awakeWithContext` function that can be used to pass an arbitrary object into the screen when it is displayed. This can be used to supply a user object, which in turn can be used to look up a set of repositories.

The first element is setting the context object when transitioning out of a screen. In the `InterfaceController` class, add a new method `contextForSegueWithIdentifier`, as follows:

```
override func contextForSegueWithIdentifier(
  segueIdentifier: String,
  inTable table: WKInterfaceTable,
  rowIndex: Int) -> AnyObject? {
   return delegate.users[rowIndex]
}
```

Now when the `RepositoryListController` is displayed, the currently-selected user will be passed through. To receive the object, create an `awakeWithContext` method in the `RepositoryListController` class, as follows:

```
override func awakeWithContext(context: AnyObject?) {
  super.awakeWithContext(context)
  if let user = context as? String {
    print("Showing user \(user)")
  }
}
```

 This will allow the code to be debugged at this point to verify that the object is being passed through as expected.

Displaying a list of repositories requires using the API to generate a list of data, creating the appropriate number of rows, and then setting the row contents as before. This can be implemented by updating the `awakeWithContext` method in the `RepositoryListController`, as follows:

```
if let user = context as? String {
  delegate.loadReposFor(user) {
    result in
    self.repositoriesTable.setNumberOfRows(
      result.count, withRowType: "repository")
    for (index, repo) in result.enumerate() {
      let controller = self.repositoriesTable
        .rowControllerAtIndex(index) as! RepositoryRowController
      controller.name.setText(repo["name"] ?? "")
    }
  }
}
```

Now when the watch application is run, and a user selected, a list of repositories should be populated in the second screen:

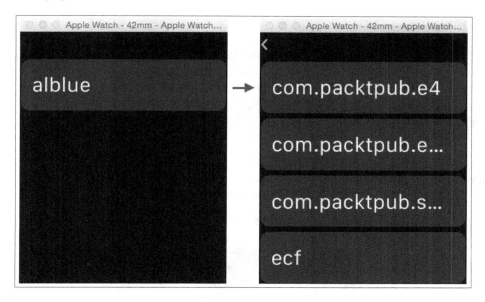

# Adding a detail screen

The final part of the watch application is to create a modal screen that is similar to the `DetailViewController` in the iOS application. When the user selects a repository, details about the repository should be presented modally.

This will be implemented with a new `RepositoryController.swift` file, which will contain a `WKInterfaceController` and have four labels that can be wired up in the interface:

```
class RepositoryController: WKInterfaceController {
  @IBOutlet weak var repo: WKInterfaceLabel!
  @IBOutlet weak var issues: WKInterfaceLabel!
  @IBOutlet weak var watchers: WKInterfaceLabel!
  @IBOutlet weak var forks: WKInterfaceLabel!
}
```

Don't forget to add the `RepositoryController.swift` file to the `RepositoryBrowserWatch Extension` target, or it will not be possible to use that as the implementation class.

To add the screen, open the `Interface.storyboard` and drag another **Interface Controller** from the object library onto the canvas. In the **Identity Inspector**, set `RepositoryController` as the **Class** type, which will allow the labels to be wired up subsequently.

Drag four **Label** objects into the watch interface. They will line up automatically in a row, one under each other. These can be given placeholder text of `Repo`, `Issues`, `Watchers`, and `Forks`—although the content of these will be changed programmatically. By dragging and dropping from the **Repository Controller** onto each of the labels, wire up the connections for the outlets so that they can be controlled programmatically.

Finally, wire up the segue from the repository list controller so that when the **repository** row controller under the **Repositories Table** is selected, a **Modal** selection segue is chosen. The completed set of connections should look like this in Xcode:

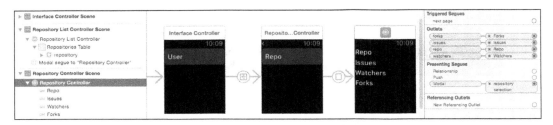

At this point the application can be tested, and selecting a repo should transition into the new screen although the correct content won't be displayed yet.

# Populating the detail screen

To wire up the labels in the detail screen, a similar process has to be followed for the previous screen: the context needs to be set from the transitioning screen, and then the data needs to be populated into the receiving screen.

In the `RepositoryListController`, the selected repository information needs to be passed on through the `contextForSegueWithIdentifier` method. However, unlike the `users` list (which is persisted in the `ExtensionDelegate`), there is no such stored repositories data list. As a result, it is necessary to persist a temporary copy of the repositories when the screen is woken.

Modify the `awakeWithContext` method of the `RepositoryListController` class to store entries in the `repos` property so that when one is selected it can be used to set the context when transitioning out of the screen:

```
var repos = []
override func awakeWithContext(context: AnyObject?) {
  super.awakeWithContext(context)
  if let user = context as? String {
    delegate.loadReposFor(user) {
      // as before
      self.repos = result
    }
  } else {
    repos = []
  }
}
override func contextForSegueWithIdentifier(
  segueIdentifier: String,
  inTable table: WKInterfaceTable,
  rowIndex: Int) -> AnyObject? {
  return repos[rowIndex]
}
```

Now when the repository is selected, the key/value pairs will be passed on through the cached content from before.

The last step in filling in the details screen is to use this context object to set up the labels. In the `RepositoryController` class, add an `awakeWithContext` method that receives the key/value dictionary, and uses the fields to display information about the repository:

```
override func awakeWithContext(context: AnyObject?) {
  if let data = context as? [String:String] {
    repo.setText(data["name"])
    issues.setText(data["open_issues_count"])
    watchers.setText(data["watchers_count"])
    forks.setText(data["forks_count"])
  }
}
```

Now when the application is run, the user should be able to step through each of the three screens to see the content.

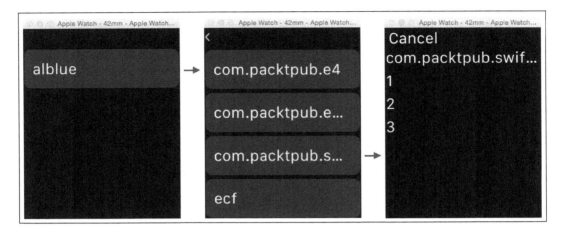

# Best practice for watch applications

As watches are very low-powered devices with limited networking, care should be taken to reduce networking where possible. The example application shown here (using several REST-based calls to a backend server) is sending and receiving more data than needed; if this was being designed as a custom application, then the protocol should be minimized to avoid unnecessary data transmission.

The example application also presented user information as a list of text data, which may not be the most appropriate way of showing data. Consider other mechanisms to present information in a more graphical way where appropriate.

## UI thread considerations

It is generally bad practice to perform any networking on the main thread, such as the lookups for the API, and for the query for a user's repositories. Instead, the lookups should be run in a background thread, switching back to the UI thread where necessary to perform updates.

For example, in the API lookup for the connection, the connect method looks like this:

```
class func connect(url:NSURL) -> GitHubAPI? {
  if let data = NSData(contentsOfURL:url) {
    ...
  }
}
```

This uses optional initializers to return a GitHubAPI whether the network connection succeeded or not, but this means that the call has to block before it can be used. This means that the GitHubAPI() initializer called in the applicationDidFinishLaunching will be blocking the application's startup, which is not excellent user experience. Instead, it is better to do something like this:

```
Threads.runOnBackgroundThread() {
  if let data = NSData(contentsOfURL:url) {
    ...
    Threads.runOnUIThread() {
      // update the UI as before
    }
  }
}
```

Adding background threads increases the complexity, but this means that the application will start faster. It may be necessary to update the UI initialization logic such that the calls to the API are deferred until the network service is available, or show other loading progress indicators to give the user feedback that something is happening.

# Stored data

The user list in the example application only stores a single variable, which is hardcoded into the application. Normally, this won't be the case, but the watch is not set up for data input. Instead, the companion iOS application should be used to define a list of users (with appropriate error checking and interface) and then communicate that with the watch application.

There are two ways of achieving this. The best way is to use the iCloud infrastructure and have the document updated on the iOS device and then mirrored to the watch automatically. This will allow the user to transition to new iOS devices or watches in the future without needing to recreate the list.

An alternative way is to send messages between the watch and the iOS device using the WatchConnectivity module and the WCSession type. This provides a singleton that is accessible through the WCSession.defaultSession(), which can be used to send and receive messages between the iOS device and the paired watch. Please note that the session may not be supported, so it should be checked with session.isSupported() first; and if it is, then it must be activated with session.activate() before any messages can be sent or received. Incoming messages are routed to the associated delegate.

The watch can also persist data using the session's `watchDirectoryURL`, which returns the location that temporary data can be written to. This can be used to add additional information which is loaded at startup. For example, the GitHubAPI could cache the API once it has been initially retrieved, then used for subsequent requests, and reloaded automatically if necessary.

# Appropriate use of complications and glances

The watch's interface predominantly uses different types of widgets for different interactions. A *complication* is a small utility widget that's displayed on the screen of the watch face (for example, the rising sun or a stopwatch timer). A *notification* is a small brief information update (similar to the notifications on iOS such as an incoming message), which can be used to perform simple actions (such as responding with a yes/no/maybe) or to launch the full application. A *glance* is a simple location-derived item that may give the user a way of telling them that something is nearby when they raise their wrist.

Depending on the type of application created, there may be appropriate ways that these can be used in order to give the user specific information on demand. However, they shouldn't be used just for the sake of using them; if they aren't going to provide any useful information, they should not be used.

There are also other ways of interacting with the application; for example, watchOS 2 has support for direct interaction of the digital crown and force pushes. For more information, see the Apple Watch Human Interface Guidelines.

# Summary

Watch applications can run code in the same way that they run on an iOS device although the way in which they are uploaded to the watch is slightly different. Running code on the simulator is very different to running on a real device; the network and processor are much more limited than will be expected for a desktop class machine (or even an iOS device). As a result, testing on a real device is essential in order to test the full experience.

This chapter presented how watch applications and extensions are built, how they are packaged in the form of watch extensions and watch apps, and how they can share code with a parent application to avoid code duplication. The watch interface demonstrated how to transition between screens using segues to implement a watch extension of the iOS application that was created in the previous chapter.

# References to Swift-related Websites, Blogs, and Notable Twitter Users

Learning any language initially focuses on the syntax and semantics of the language, but it quickly moves on to learning the suite of both standard and additional libraries that allow programmers to be productive. A single book cannot hope to list all possible libraries that will be needed; this book is intended to be the start of a learning journey.

For further reading, this appendix presents a number of additional resources that may be useful to the reader in order to continue this journey. In addition, look out for other books by Packt Publishing that present different aspects of Swift. This list of resources is necessarily incomplete; new resources will become available after the publication of this book, but you may be able to find new developments as they occur by following the feeds and posts of the resources given here.

## Language

The Swift language is developed by Apple, and a number of documents are available from the Swift developer page at `https://developer.apple.com/swift/`. This includes a language reference guide and an introduction to the standard library:

- The Swift programming language can be found at `https://developer.apple.com/library/ios/documentation/Swift/Conceptual/Swift_Programming_Language/`

- The Swift standard library reference can be found at `https://developer.apple.com/library/ios/documentation/General/Reference/SwiftStandardLibraryReference/`
- Integrating Swift and Cocoa can be found at `https://developer.apple.com/library/ios/documentation/Swift/Conceptual/BuildingCocoaApps/`
- Swifter provides a list of all Swift functions at `http://swifter.natecook.com`

The Swift language was open sourced in December 2015 and has a new home at `https://swift.org`, along with the new Swift blog at `https://swift.org/blog/`.

# Twitter users

There are a lot of active Twitter users that use Swift; in many cases posts will be marked with the `#swift` hashtag, and can be found at `http://twitter.com/search?q=%23swift`. Popular users that the author follows include (in alphabetical Twitter handle name):

- `@AirspeedSwift`: This twitter has a good selection of tweets and retweets of Swift-related subjects
- `@ChrisEidhof`: This is author of the *Functional Swift* book and `@objcio`
- `@CodeWithChris`: This twitter is a collection of tutorials on iOS programming
- `@CodingInSwift`: This twitter contains cross-posts by a collection of Swift resources
- `@CompileSwift`: This twitter contains posts on Swift
- `@cwagdev`: Chris Wagner writes some of the iOS tutorials with Ray Wenderlich
- `@FunctionalSwift`: This is a selection of functional snippets, along with a Functional Swift book
- `@LucasDerraugh`: This is the creator of video tutorials on YouTube
- `@NatashaTheRobot`: This twitter contains a great summary of what's happening, along with newsletters and cross references
- `@nnnnnnnn`: Nate Cook, who reviewed an earlier version of this book and provides the Swifter list just mentioned
- `@PracticalSwift`: This is a good collection of blog posts talking about the Swift language
- `@rwenderlich`: Ray Wenderlich has many posts relating to iOS development; a wealth of information and more recently Swift topics as well

- @SketchyTech: This is a collection of blog posts on Swift
- @SwiftCastTV: These are video tutorials of Swift
- @SwiftEssentials: This is the twitter feed for this book
- @SwiftLDN: This Twitter posts Swift meetups based in London, also invites great Swift talks and presenters

In addition to the Swift-focused Twitter users, there are a number of other Cocoa (Objective-C) developers who blog regularly on topics relating to the iOS and OS X platforms. Given that any Objective-C framework can be integrated into a Swift app (and vice versa), quite often, there will be useful information from reading these posts:

- @Cocoanetics: Oliver Drobnik writes about iOS and provides training
- @CocoaPods: CocoaPods is a dependency management system for Objective-C frameworks (pods) and is being extended into the Swift domain
- @Mattt: Mattt Thompson writes about many iOS subjects, is the author of the AFNetworking and AlamoFire networking libraries, and who moved to Apple to write the Swift package manager
- @MikeAbdullah: Mike Abdullah writes about general iOS development
- @MikeAsh: Mike Ash knows everything there is to know, and what he doesn't know, he finds out
- @MZarra: Marcus S. Zarra has written a lot about Core Data and synching
- @NSHipster: This is a collection of assembled iOS and Cocoa posts that are organised by Mattt Thompson
- @objcio: This is a monthly publication on Objective-C topics with some Swift
- @PerlMunger: Matt Long posts about Swift, Cocoa, and iOS

The reviewers of this book included:

- @AnilVrgs: Anil Varghese
- @Ant_Bello: Antonio Bello
- @ArvidGerstmann: Arvid Gerstmann
- @jiaaro: James Robert
- @nnnnnnnn: Nate Cook

The author's personal and book twitter accounts are:

- @AlBlue is the author's twitter account
- @SwiftEssentials is the book's twitter account

Meetups such as @SwiftLdn keep a track of interesting Swift writers in a Twitter list at https://twitter.com/SwiftLDN/lists/swift-writers/members, which may have more up-to-date recommendations than this section, as well as the Ray Wenderlich team at https://twitter.com/rwenderlich/lists/raywenderlich-com-team/members.

# Blogs and tutorial sites

There are a number of blogs that cover Swift and related technologies. Here are a selection that you may be interested in:

- https://developer.apple.com/swift/blog/ is the official Apple Swift blog
- http://airspeedvelocity.net is the blog for @AirspeedSwift
- http://alblue.bandlem.com/Tag/swift/ is the author's blog on Swift
- http://mikeabdullah.net is Mike Abdullah's blog
- http://mikeash.com writes the Friday Q&A series on all things iOS and OS X
- http://natecook.com/blog/tags/swift/ is Nate Cook's blog on Swift
- http://nshipster.com is the blog for @NSHipster
- http://objc.io is the blog for @objcio
- http://practicalswift.com is collected by @PracticalSwift
- http://sketchytech.blogspot.co.uk is a collected blog of Swift articles by @SketychTech
- http://swiftessentials.org is the companion site for this book, along with the repository at https://github.com/alblue/com.packtpub.swift.essentials/
- http://swiftnews.curated.co is collected by @NatashaTheRobot
- http://www.cimgf.com presents a collection of topics on Cocoa, by Marcus S Zarra and others
- http://www.raywenderlich.com has a collection of tutorials about iOS development, including both Cocoa and Swift

# Meetups

A number of local iOS developer groups existed before Swift was created; they have since been supplanted by Swift-specific groups. These will of course vary by geographic location, but a few meetup sites exist, such as EventBrite at http://www.eventbrite.co.uk, and Meetup at http://www.meetup.com.

There are also likely to be Twitter groups or meetups near you; for example, in London, there is `@SwiftLDN` at `https://twitter.com/SwiftLDN` who have regular meetings listed at `http://www.meetup.com/swiftlondon/`. In New York, the `http://www.meetup.com/NYC-Swift-Developers/` group is fairly active. In San Francisco, both `http://www.meetup.com/swift-language/` and `http://www.meetup.com/San-Francisco-SWIFT-developers/` are active.

# Afterword

*A journey of a thousand miles begins with a single step.* Your journey to writing great Swift applications has just begun. As with any journey, traveling companions can provide support, assistance, and encouragement; and many of the companions given here can provide connections to many more. I hope you enjoy your journey.

# Index

# V

variables 5

# W

**watch application**
  about 193
  best practice 210
  GitHubAPI, adding to
        watch target 196, 197
  watch target, adding 194, 195
**watch interfaces**
  creating 197
  image, adding 202
  list of users, adding to watch 197, 198
  wiring up 199-201

# X

**XCPlayground framework**
  about 42
  asynchronous code, executing 44, 45
  values, capturing explicitly 42-44
**XCTest framework 65**
**XML**
  data, downloading 148
  data, parsing 149, 150
  parser delegate, creating 148
  parsing 147

## Thank you for buying
# Swift Essentials
### Second Edition

# About Packt Publishing

Packt, pronounced 'packed', published its first book, *Mastering phpMyAdmin for Effective MySQL Management*, in April 2004, and subsequently continued to specialize in publishing highly focused books on specific technologies and solutions.

Our books and publications share the experiences of your fellow IT professionals in adapting and customizing today's systems, applications, and frameworks. Our solution-based books give you the knowledge and power to customize the software and technologies you're using to get the job done. Packt books are more specific and less general than the IT books you have seen in the past. Our unique business model allows us to bring you more focused information, giving you more of what you need to know, and less of what you don't.

Packt is a modern yet unique publishing company that focuses on producing quality, cutting-edge books for communities of developers, administrators, and newbies alike. For more information, please visit our website at www.packtpub.com.

# About Packt Open Source

In 2010, Packt launched two new brands, Packt Open Source and Packt Enterprise, in order to continue its focus on specialization. This book is part of the Packt Open Source brand, home to books published on software built around open source licenses, and offering information to anybody from advanced developers to budding web designers. The Open Source brand also runs Packt's Open Source Royalty Scheme, by which Packt gives a royalty to each open source project about whose software a book is sold.

# Writing for Packt

We welcome all inquiries from people who are interested in authoring. Book proposals should be sent to author@packtpub.com. If your book idea is still at an early stage and you would like to discuss it first before writing a formal book proposal, then please contact us; one of our commissioning editors will get in touch with you.

We're not just looking for published authors; if you have strong technical skills but no writing experience, our experienced editors can help you develop a writing career, or simply get some additional reward for your expertise.

**open source**
community experience distilled

[PACKT] PUBLISHING

# Swift 2 Design Patterns

ISBN: 978-1-78588-761-1          Paperback: 224 pages

Build robust and scalable iOS and Mac OS X game applications

1. Learn to use and implement the 23 Gang of Four design patterns using Swift 2.

2. Design and architect your code for Swift application development.

3. Understand the role, generic UML design, and participants in the class diagram of the pattern by implementing them in a step-by-step approach.

# Swift 2 Blueprints

ISBN: 978-1-78398-076-5          Paperback: 276 pages

Sharpen your skills in Swift by designing and deploying seven fully-functional applications

1. Develop a variety of iOS-compatible applications that range from health and fitness to utilities using this project-based handbook.

2. Discover ways to make the best use of the latest features in Swift to build on a wide array of applications.

3. Follow step-by-step instructions to create Swift apps oriented for the real world.

Please check **www.PacktPub.com** for information on our titles

www.ingramcontent.com/pod-product-compliance
Lightning Source LLC
Chambersburg PA
CBHW060543060326
40690CB00017B/3583